Active Ageing in the European Union

Active Ageing in the European Union

Policy Convergence and Divergence

Kate A. Hamblin
Oxford Institute of Population Ageing, University of Oxford, UK

palgrave
macmillan

First published 2013 by
PALGRAVE MACMILLAN

Palgrave Macmillan in the UK is an imprint of Macmillan Publishers Limited, registered in England, company number 785998, of Houndmills, Basingstoke, Hampshire RG21 6XS.

Palgrave Macmillan in the US is a division of St Martin's Press LLC, 175 Fifth Avenue, New York, NY 10010.

Palgrave Macmillan is the global academic imprint of the above companies and has companies and representatives throughout the world.

Palgrave® and Macmillan® are registered trademarks in the United States, the United Kingdom, Europe and other countries

ISBN 978-0-230-35370-1

This book is printed on paper suitable for recycling and made from fully managed and sustained forest sources. Logging, pulping and manufacturing processes are expected to conform to the environmental regulations of the country of origin.

A catalogue record for this book is available from the British Library.

A catalog record for this book is available from the Library of Congress.

Transferred to Digital Printing in 2013

Contents

List of Figures and Tables

Figures

Tables

Acknowledgements

The research discussed in this book was funded by the Economic and Social Research Council, research grant number PTA031200400138. Their support is gratefully acknowledged. My thanks also go to Theo Papado-poulos for his support and supervision and Emma Carmel for providing inspirational teaching during my studies as an undergraduate and postgraduate student, as well as excellent advice. I'd also like to thank Professor Bleddyn Davies for his comments on draft sections of this book.

List of Abbreviations

ABW	*Algemene Bijstandswet* (Law on General Social Assistance (Netherlands))
ACA	*Allocation Chômeurs Âgés* (Older Unemployed Compensation (France))
ACC	Solidarity gradual preretirement contracts (France)
ACS	*Allocation conventionnelle de solidarité*
ADE	*Aide dé à l'employeur* (Degressive Employer's Subsidy (France))
AER	*Allocation équivalent retraite* (Retirement Equivalent Benefit (France))
AET	Adult Education and Training Courses
AGIRC	*Association Générale des Institutions de Retraite des Cadres* (General Association of Retirement Institutions for Executives (France))
ALMP	Active labour market policies
AMS	*Arbeitsmarktservice* (Austrian public employment service)
AMFG	*Arbeitsmarktfördesrungsgetz* (Austrian Labour Market Promotion Act)
AMSG	*Arbeitsmarktservicegesetz* (Austrian Public Employment Service Act)
AOW	*Algemene Ouderdoms Wet* (General Seniority Law (Netherlands))
ARPE	*L'Allocation de Remplacement Pour l'Emploi* (Combined early retirement/hiring scheme, also known as *mise à la retraite* (France))
ARRCO	*Association pour le Régime de Retraite Complémentaire des salaries* (Private Association of Complementary Retirement Systems (France))
ASA	*Allocation Spécifique d'Attente* (specific allocation of waiting (France))
ASFNE	*Allocation spéciale du Fonds national de l'emploi* (Special National Employment Fund Allowance (France))
ASS	*Allocation de solidarité spécifique* (Specific Assistance Allowance (France))
ASSEDIC	*Association pour l'emploi dans l'industrie et le commerce* (Unemployment Insurance Fund (France))
ASVG	*Allgemeines Sozialversicherungsgesetz* (Austrian General Social Insurance Act)

ATP	*Arbejdsmarkedets tillægspension* (Supplementary Pension (Denmark))/*tilläggspension* (Supplementary Pension (Sweden))
CAATA	*Cessation Anticipée d'Activité des Travailleurs de l'Amiant* (Early withdrawal for workers who were exposed to asbestos (France))
CAE	*Contrats d'accompagnement dans l'emploi* (Accompanying Contracts of Employment (France))
CASA	*Cessation d'activité des salariés âgés* (Early Retirement for Older Workers (France))
CATS	*Certains Travailleurs Salariés* (early retirement scheme for employees (France))
CEC	Commission of the European Communities
CEOC	Spanish Confederation of Employers' Organisations
CES	*Contrat Emploi Solidarité* (Employment solidarity contract (France))
CDD	*Contrat Dernière Embauche* (Last hiring contract (France))
CEU	Council of the European Union
CFA	*Congé de Fin d'Activitié* (end of employment leave (France))
CIE	*Contrat d'Initiative Emploi* (Contract employment initiative (France))
CIG	*Cassa Integrazione Guadagni* (wage guarantee fund (Italy))
COR	Committee of the Regions
CPA	*Cessation Progressive d'Activité* (Phased-in retirement (France))
CRE	*Contrat de Retour à l'Emploi* (Contract back-to-work (France))
DIF	*Droit Individuel à la Formation* (individual right to training (France))
DKK	Danish Krone
DRA	Default Retirement Age
DRE	*Dispense de Recherche d'Emploi* (Exemption from job search (France))
DTI	Department of Trade and Industry
DWP	Department of Work and Pensions
EC	European Commission
EES	European Employment Strategy
EU	European Union
FAS	Foras Áiseanna Saothair (Training and Employment Authority (Ireland))
FNE	*Fonds National pour l'Emploi* (National Employment Fund (France))

GDR	German Democratic Republic
GRD	*Garantie de Resources Demission* (France)
GRL	*Garantie de Resources Licenciement* (France)
HALDE	High Authority to Combat Discrimination and Promote Equality (France)
ICT	Information and Communication Technology
IESG	Insolvency Contingency Fund (Austria)
IKA	Social Insurance Institute (Greece)
ILO	International Labour Organisation
IPOS	International Organisation of Pension Supervisors
KWNS	Keynesian Welfare National State
LAEK	Employment and Vocational Training Fund (Greece)
MISSOC	Mutual Information System on Social Protection
MISSPEU	Mutual Information System on Social Protection in the European Union
NAP	National Action Plan
ND50+	New Deal 50 Plus (UK)
OAED	Manpower Employment Organisation (Greece)
OECD	Organisation for Economic Cooperation and Development
OMC	Open Method of Co-ordination
OTA	*Offentliga Tillfälliga Anställningar* (Public Temporary Employment (Sweden))
PAP	*Plan d'action personnalisé* (Personalised Action Plan (France))
PAP-ND	*Programme d'Action Personnalisé pour un Nouveau Depart* (Personalised Action Plan for a New Start (France))
PARE	*Plan d'Aide au Retour à l'Emploi* (return-to-work action plan (France))
PAYG	Pay-as-you-go
PES	Public Employment Service
PPI	Pensions Policy Institute
PRETA	Pre-Retirement Allowance (Ireland)
PRP	*Préretraite Progressive* (gradual retirement (France))
PRSI	Pay Related Social Insurance (Ireland)
PVK	Pension and Insurance Supervisory Authority
RGAVTS	*Régime général d'assurance vieillesse des travailleurs salariés* (General Scheme for Employees (France))
RMI	*Revenu Minimum d'Insertion* (minimum income support allocations (France))
SEK	Swedish Krona
SWPR	Schumpeterian Workfare Postnational Regime

SZW	*Ministerie van Sociale Zaken en Werkgelegenheid* (Ministry of Social and Employment Affairs)
TLM	Transitional Labour Market Theory
TUC	Trades Union Congress
UA	Unemployment Assistance
UI	Unemployment Insurance
UNEDIC	*Union Nationale pour l'Emploi dans l'Industrie et le Commerce* (Unemployment administration (France))
VAE	*Validation des Acquis de l'Expérience* (Validation of Learning Through Experience (France))
VERP	*Efterløn* (Voluntary Early Retirement Pension (Denmark))
VUT	*Vervroegde Uittreding* (early retirement pension (Netherlands))
WAO	*Wet op de Arbeidsongeschiktheidsverzekering* (occupational disability programme (Netherlands))
WGA	Work Resumption Benefit for Persons Partially Capable of Work
WGBL	Dutch Equal Treatment Act on the Basis of Age in Employment
WHO	World Health Organization
WIA	*Werk en Inkomen naar Arbeidsvermogen* (Work and Income (Capacity for Work) Act (Netherlands))
WIW	*Wet inschakeling werkzoekenden* (Law on the Retaining of Job-seekers (Netherlands))
WW	*Werkloosheidswet* (unemployment benefits (Netherlands))

1
Introduction

Increased longevity and declining fertility are prompting concerns about the sustainability of current welfare arrangements. 'Active ageing' policies have been presented as a potential means to arrest the conflict many argue will be produced by the increased share of the older population and the corresponding demands which will be placed on pension and care systems. Though 'active' or 'productive' ageing are not new concepts, this renewed focus is also argued to be part of a broader shift away from 'passive' to 'active' social policies. When initially created, activation and active welfare policies focused on young unemployed individuals, but increasingly groups of formerly 'deserving' welfare recipients are now the targets of these policies, including for example older individuals,[1] lone parents and those with disabilities.

However, 'active ageing' is a contested concept, with organisations such as the World Health Organization (WHO) focusing on broad elements which enhance the wellbeing of older individuals. This organisation's active ageing model includes three pillars: health, participation and security. On the other hand, organisations such as the EU and World Bank concentrate more specifically on 'activity' in older age in terms of labour market participation for the most part. Indeed, the EU has created two targets for its member states which encourage the re-integration and retention of those aged 55–64 in the labour market.[2] In focusing on EU15 nations, this book will also concentrate on the EU's definition of 'active ageing', though alternative conceptualisations will also be discussed in Chapter 2.

This book will address the extent to which EU15 nations' 'active ageing' strategies follow the EU's model and will therefore focus on policies for employment and retirement for older people in three ways.[3] First, in order to explore the degree to which they emulate the EU-vision of active ageing, empirical data on national policy pictures will be explored to

1

identify general convergence, the formation of clusters, or distinct national approaches with regard to pensions and unemployment policies for older individuals in EU15 nations from 1995–2010. Second, the character of these reforms over the fifteen-year period will be explored, reflecting the contribution of the new institutionalist literature (which will be addressed in Section 3.2; c.f. Pierson, 2001, 2004) and its focus on national divergence and the importance of policy legacies. Finally the book will address which sub-groups within the category of 'older age' are subject to the retrenchment of early exit routes and new policies aimed at retaining and re-integrating individuals into the labour market. As emphasised by the political economy of ageing literature (explored further in Section 3.6), the experience of ageing is not homogenous and social policy plays a role in its construction.

In terms of the book's structure, the first substantive chapter will address issues around ageing, activity and passivity. For Walker (1996a: 2), at the macro-level, intergenerational solidarity has been formalised by welfare state arrangements in that transfers between generations have 'encouraged the expectation of reciprocity'. However, as a result of the shift away from *decommodifying* welfare arrangements (i.e. social policies which provided an alternative to the sale of labour on the market) towards the *recommodification* of labour (i.e. social policies which promote or insist upon the sale of labour on the market) and the active ageing agenda, the contract between the generations is being recalibrated, with older individuals encouraged to secure their own welfare, either by engaging in paid labour or through purchase on the market (Carmel et al., 2007).

Chapter 3 will narrow the focus to explore the EU's active ageing agenda, and outlines the approach this volume will take when examining EU15 nations' policies for older individuals. This chapter also addresses the literature on policy divergence, which acts as a caveat to avoid the overemphasis on policy convergence between EU15 nations. Indeed, the EU has advocated an active ageing agenda that focuses on labour market participation yet the rate, type and target groups within the older age cohort vary in different member states. The policy legacies of EU15 nations provide different environments for active ageing policies and subsequent reforms. Indeed, there was variation in the degree to which the decommodification of this age group was previously encouraged by policy, which meant that some EU15 nations had greater distances to travel in order to implement the EU's active ageing model. In addition, the older age cohort is not a homogenous group, interacting with social policy in the same way in terms of inputs and outcomes. The political economy of ageing literature emphasises the role social structures, particularly the state, have on the experience of ageing. Though ageing is a biological fact, many of the social facets are constructions and are experienced differently by the various sub-groups that make up a particular cohort. The experience of ageing is filtered through other individual characteristics, such as gender, class and ethnicity, and differential treatment according to these characteristics is carried through the lifecourse (Carr and Sheridan, 2001; Estes, 1991; Heinz, 2001; Phillipson, 2005). Particularly pertinent to this publication is the notion that the experience of ageing is bound to the individual's labour market history, and therefore disadvantage and discrimination faced by different groups during their working lives are transferred into old age. Thus the shift towards the recommodification of labour is not universally experienced by all 'older' individuals, as demonstrated by the data included in subsequent chapters.

In 2004, the EU classified EU15 nations in accordance with the position around the Stockholm target of 50% employment for the 55–64 age group in its *Draft Joint Employment Report 2003/2004* (Council of the European Union, 2004: 8). This classification grouped nations into those who had reached the 50% target or were near in 2002 (Sweden, Denmark, the UK and Ireland), those which were 'particularly worrisome' as their participation rate for older people was less than 35% (Belgium, Luxembourg, Italy, Austria and France), and an 'inbetween group' (Germany, Finland, Spain, the Netherlands and Greece) which was close to the EU average. In Chapters 4 to 7, this volume explores the policies available for older individuals in EU15 nations and revisits this classification to focus on where EU15 nations were in relation to this target by the 2010 deadline. Taking into account the differences in participation rates in 2001 and 2010, the following typology was created which considers both on nations' overall progress towards the Stockholm target and the distance they had to travel to reach 50% employment for 55–64 year olds:

– *Group I – The Vanguards:* Nations that were in 2001 above or close to the Stockholm target and have since maintained or strengthened this position, including Sweden, Denmark, the UK, Ireland and Portugal.
– *Group II – Surpassing Stockholm:* Nations that since 2001 have moved beyond the Stockholm target including the Netherlands, Finland and Germany.
– *Group III – Below Stockholm but approaching fast*: Nations that have not yet met the Stockholm target, but have made progress beyond the average percentage change for EU15 nations between 2001–2010 (9.6%). These nations include Belgium, Austria and Luxembourg.

- *Group IV – The Laggards*: Nations that have not yet met the Stockholm target and whose progress was less than the average for EU15 nations (9.6%), including France, Greece, Italy and Spain.

In each chapter, the countries' interventions for older individuals are examined in terms of their labour market (unemployment benefit extensions and job search exemptions – which represent *de facto* early exit policies – and active labour market schemes) and pension policies (pension principles, early retirement schemes, incentives for deferring pension receipt). In addition, in order to explore the potential differences in policies available at the micro-level, 'model biographies' were used to address the choices ideal-typical older individuals would face in each nation, and how these were altered over time. The model biography approach (influenced by Bradshaw et al. (1993) and Meyer et al. (2007)) allowed for the exploration of the impact individual characteristics such as employment history, age and gender have on individual choice regarding labour market participation and exit. The model biographies utilised were as follows:

- the 50-plus:
  - *Laurent:* Aged 55 with 35 years of employment contributions.
  - *Jean:* Aged 55 with a disjointed employment history.
- the 60-plus:
  - *Sasha:* Aged 63 years with 35 years of employment contributions.
  - *Jude:* Aged 63 with a disjointed employment history.

The data clearly indicate that, as suggested by the political economy of ageing literature, state policies do not interact with all older people in the same way as policy treatment is determined by individual characteristics which then in turn influence the experience of ageing. It became clear when exploring the policy options for the different model biographies that age was an important division between individuals within the older age cohort in terms of their ability to be decommodified and exit, or remain in/re-enter the labour market and be recommodified. Often policies for decommodification did not directly focus on the 50–55 age cohort from the outset, or the policy picture remained static over the ten-year period. It was noted however, that these nations did provide active labour market policies (ALMPs) that were applicable to *Laurent* and *Jean* and perhaps in this way, there was not so much a shift from decommodifying welfare arrangements, but an increased onus on recommodification. In addition, it is apparent that gender too was important in that in some nations' age thresholds were not harmonised, allowing women to exit earlier than their male counterparts. However, in other nations, it becomes clear that though age and gender provide eligibility parameters, access to decommodification was contingent upon other factors such as contribution records or the labour market situation.

Thus the policy approaches of EU15 nations can be divided into those where decommodification was dependent upon desert through contributions (and thus presented *Laurent* and *Sasha* with more opportunities for exit due to their 35 years of contributions); those where decommodification was dependent upon the labour market situation (and therefore provided exit according to occupation or economic situation, thereby treating all model biographies the same in certain contexts); and those which largely provided rights-based decommodification options (where the policy packages for *Laurent, Jean, Sasha* and *Jude* would have been identical, irrespective of contribution records) (see Table 1.1 below). As with any categorisation, a completely perfect fit not always possible yet the within-group differences are considerably less than those between groups. Indeed as Arts and Gelissen (2002: 140) note "[a]lthough real welfare states are most of the time not unique, they certainly are never completely similar. This means that they are almost always impure types. The consequence is that although they cluster together in three subclasses it is not always easy to classify all cases unambiguously".

Table 1.1 Typology of decommodification approaches

Decommodification eligibility focus	Early exit/retirement policy characteristics
Rights	Early exit/retirement schemes open to all over a certain age.
Desert	Access to early exit/retirement schemes conditional on previous contribution record.
Labour market	Access to early exit/retirement schemes conditional on employment in a particular industry/particular type of work/with a 'replacement' condition.

Chapter 8 provides an analytical discussion of the data from Chapters 4 to 7 with reference to the literature from the first two chapters. Though the data indicate that EU15 nations' policies for older people from the mid-1990s to 2010 have moved towards 'active ageing' insofar as early exit and retirement has been retrenched and activation policies have been introduced, there are two important caveats. First, at the macro-level the picture is more complex than linear convergence with nations moving towards active ageing at different speeds, with different policy mixes and legacies to contend with. Second, at the micro-level, the policy options available to the 'older' age cohort are not necessary uniform and thus some groups are encouraged to age more actively than others.

Finally, Chapter 9 concludes that the data demonstrate that though EU15 nations have adopted elements of the EU15's active ageing approach, there is a great deal of variation in the type and timing of reforms, in part influenced by policy legacies and contexts. Also, there is divergence both between and within nations with regard to their treatment of the older age cohort; not all individuals are equally subject to the active ageing agenda. Thus though the EU discourse may advocate these policies in the hope of preserving intergenerational solidarity, they in turn have an impact on intragenerational equity.

2
Active Ageing: Origins and Resurgence

This book explores the recent pension and unemployment policy development-ments for older people in EU15 nations with a view to determining whether there has been a shift towards the EU's 'active ageing' agenda in all nations and for all sub-groups within the 'older age' cohort. As such it begins by examining the development of the concept of 'active ageing' before presenting arguments regarding its resurgence in the 1990s. It therefore discusses the demographic pressures faced by current welfare arrangements and the potential intergenerational conflict that may arise as a result, as well as situating the active ageing agenda within the broader shifts towards either the recommodification of labour or the inclusion of older individuals in the 'reserve army of labour'. Though this chapter will explore some of the potential explanations for focus on 'active ageing', it seeks to provide an introduction to these debates as opposed to a definitive answer or identify a causal relationship; this book instead aims to address the extent to which EU15 nations have subscribed to the EU's vision of policies for older individuals.

2.1 The move away from 'passivity': Active ageing

Active ageing is a relatively new term, becoming gradually more salient in the last ten years. It first originated in the United States of America in the 1960s with literature on 'successful ageing' which focused on "denying the onset of old age and by replacing those relationships, activities and roles of middle age that are lost with new ones in order to maintain activities and life satisfaction" (Walker, 2002: 122). This perspective emerged as a response to disengagement theory, which argued as individuals age, they gradually withdraw from many social spheres and reduce the roles they take in society (Bond et al., 1993 in

Avramov, 2003; Cumming and Henry, 1961). Activity theorists countered this approach, arguing it presented an overly depressing picture
of older age and maintained that when faced with a decline in their
traditional roles, ageing individuals engage in new activities in order to
compensate for this loss (Havighurst et al., 1968 in Avramov, 2003).

The concept of active ageing underwent a renaissance in the 1980s
under the guise of 'productive ageing' and shifted its focus to include
the whole lifecourse as opposed to the latter stages (Walker, 2002). The
new emphasis on the concept at this time was driven by the desire on
the part of individuals to get more from their years in retirement and
policy makers' concerns at the costs of ageing on state provision. As
a result this approach took a narrow, economistic perspective, with
productive ageing defined as "any activity by an older individual that
produce goods and services, or develops the capacity to produce goods
or services" (Caro et al., 1993: 6).

The evolution of productive ageing into 'active ageing' occurred
in the 1990s with the World Health Organization (WHO) which argued
there was a link between activity and health (Walker, 2002). The WHO[1]
(2002) takes a holistic approach to active ageing, defined as "the process
of optimizing opportunities for health, participation and security in order
to enhance quality of life as people age" and acknowledges citizens' rights,
needs and preferences. Thus it represented the move from a needs-based
approach to one focused more strongly on rights, whilst at the same time
stressing these are not devoid of responsibilities. The WHO emphasises
the importance of quality of life in younger years for determining wellbeing in later life, as well as the social nature of ageing, in that the process
does not occur in isolation thus making "interdependence as well as
intergenerational solidarity (two-way giving and receiving between individuals as well as older and younger generations)... important tenets of
active ageing" (ibid: 12). As a result the WHO definition defines activity
as including "continuing participation in social, economic, cultural, spirit-
ual and civic affairs, not just the ability to be physically active or to participate in the labour force" (ibid: 12) and has three pillars of active ageing:
health, participation and security.

The approach of the WHO contrasts with the 'productive ageing' discourse of the 1980s which has re-emerged in the way supranational
organisations such as the EU and World Bank construct the term active
ageing.[2] Therefore a split can be seen between holistic approaches to
active ageing, as exemplified by the WHO, and those organisations which
focus on labour market participation. The latter present a form of active
ageing which focuses on aspects such as preventative healthcare and life-

long learning as the *means* to encourage older individuals to re-enter or remain in paid employment. Avramov and Maskova (2003: 24) for the Council of Europe make the distinction between the "active way of spending the increased free time after retirement (e.g. World Health Organization) others are mainly interested in economic activity as labour force participation (e.g. OECD; European Commission). In recent years a shift can be perceived from the first towards the latter preoccupation". However, the portrayal of retirement as purely a period of passivity because it does not involve labour market participation underestimates other social roles occupied by older individuals. Retirement does not necessarily indicate inactivity; often it is a period characterised by voluntary work or care for grandchildren (Guillemard and Rein, 1993; Künemund and Kolland, 2007). Laslett (in Hunt, 2005) also argues that though a group may not be productive in the traditional sense of labour market participation, they are still consumers and thus retain an important role for societal functioning. This book will return to the EU's conception of active ageing in Chapter 3; first the remainder of this chapter will explore the developments behind the recent resurgence of the concept.

2.2   The commodification, decommodification and recommodification of labour

Older individuals are increasingly encouraged to remain in the labour market as part of the EU's 'active ageing' agenda. However, this shift towards 'active ageing' can be seen as part of a broader move from 'passive' to 'active' welfare arrangements which represent the 'recommodification of labour'. The transition away from decommodifying social policies has implications for the relationships at the heart of welfare arrangements and the position of older individuals within them. Whereas previous passive welfare arrangements allowed for the 'decommodification of labour'[3] in that the individual could survive without recourse to the sale of their labour on the market, the focus on workfare, activation and employment has increased. However, the degree of decommodification provided was not absolute and varied nationally: "it is not all or nothing [...] the concept refers to the degree to which individuals, or families, can uphold a socially acceptable standard of living independently of market participation" (Esping-Andersen, 1990: 37). Thus the variation of welfare state arrangements results in different levels of decommodification, as will be explored in later chapters which address early exit and retirement policies and their eligibility criteria.

Jessop (1994, 1998, 1999, 2002) argues exogenous and endogenous pressures have prompted this shift within welfare arrangements whereby social policy has become the means by which workers and the labour market are made more flexible in line with the needs of the knowledge-based economy, which in turn has implications for the construction of labour as a 'fictitious commodity' (c.f. Polanyi, 1944). Jessop (1999 and 2002) argues there has been a shift from Keynesian Welfare National States (KWNS) to a Schumpeterian Workfare Postnational Regime (SWPR) which represents the move from decommodifying welfare arrangement towards the recommodification of labour through activation and workfare policies.

Jessop (2002) argues the 'passive' welfare policies were introduced as a solution to the problems of a different era; the political, economic and social foundations upon which welfare arrangements were built shifted, exposing structural weaknesses, leaving it open to criticism and restructuring. Jessop's work emphasises inherent 'crisis tendencies' within these arrangements which have contributed to the shift, such as its expansionary dynamic but does not neglect the external changes that have prompted the move away from 'passive' decommodifying welfare arrangements, including the ageing population, which will be explored in more detail in Section 2.3.

The rise of 'active' welfare policies realign the focus of the welfare state from providing for need outside of the labour market to encouraging the re-engagement with paid employment through a mixture of sanctions and incentives and as such can be argued to represent the 'recommodification of labour'. Though it is the case that decommodifying benefits were often conditional on the individual's relationship with the labour market through contribution requirements, activation policies instead shift the focus from *previous* engagement in paid employment to make the receipt of benefits conditional on *current or future* labour market behaviour (Carmel and Papadopoulos, 2003), i.e. whether the individual agrees to engage in certain activities such as job search, training, work placements etc.

Active welfare policies do however encompass a range of approaches and elements such as 'workfare', 'activation' and 'welfare-to-work' which some authors argue differ in their approach. Torfing (1999) argues there is in actuality a continuum from workfare to welfare, with activation closer to the former. Torfing makes the distinction between offensive unemployment strategies that deal with economic difficulties in a "proactive manner... to produce a positive-sum solution" (ibid: 9) and defensive strategies, which aim for "zero-sum solutions that give in to short-term partisan interests or ideological concerns". Therefore 'workfare' denotes 'defensive' work-focused programmes, whilst 'activation' refers to 'offensive' schemes with an equal focus on training. Yet, as Quaid (2002: 19) argues "conceptual distinctions exist between the terms... these differences are largely semantic and reflect more the preference of the politicians for how the words might sound to the voter's or taxpayer's ear than any actual, tangible distinctions between types of programs. Basically, workfare signifies a form of welfare for which recipients undertake some labour-market-related activities... in return for government payment". Perhaps then though their means differ, as the aim of workfare, welfare-to-work and activation policies is to promote labour market participation, they should both be included as 'recommodifying' welfare arrangements. Indeed, all seek to move recipients towards the labour market, away from decommodification.

Dean (2007) creates a typology of welfare-to-work regimes, which incorporates both workfare and activation. Underlying welfare-to-work policy approaches, Dean argues, is the nation's commitment to either solidaristic or contractarian notions of citizenship. In terms of the former, the welfare-to-work regimes aim to increase the social inclusion of the individual, whilst the latter binds the individual in a contract with the state designed to increase labour market competitiveness. Intersecting this dichotomy is the moral premise underlying the welfare-to-work approaches: whether they focus on fairness or are authoritarian and thus concerned with social order.

Using the way welfare-to-work or workfare policies balance these two dichotomies, Dean then divided them into a fourfold taxonomy. In terms of those types of workfarist policies which are constructed around concerns competitiveness whilst striving for equality and inclusion, Dean argues 'human capital' is key in that it "focuses on the opportunities that are made available to individuals to enhance their productive potential and their labour market readiness" and so includes training and education measures (Dean, 2007: 583). In this system, the individual is responsible for ensuring they are sufficiently skilled so as not to become a burden on the state through benefit receipt. Policies balancing concerns regarding social order and promoting competitiveness embody a 'work first' approach, utilising 'defensive' workfarist policies which do not take into account the quality or appropriateness of the employment; the onus on the individual to enter paid employment, irrespective of personal cost. Dean names this approach 'coercive/work first'. In those systems striving for equality, fairness and inclusion with their unemployment policies, work is integral to societal functioning and individual integration and as a result policies are distributed universally. Dean argues in this system, as embodied by Scandinavian nations until the late 20th century, featured the state as the employer in the first instance though these policies have now shifted towards subsidies and increasing employment opportunities in the private sector and classifies them as 'active job creation'. Finally, where there is a balance between inclusiveness and maintaining social order and individual contracts with the state, "welfare-to-work is concerned to promote the 'insertion' of those who have been excluded into the labour market; to give effect to their right and their moral obligation to work" (Dean, 2007: 584).

Bonoli (2010) also constructs a similar taxonomy which includes 'incentive reinforcement', akin to Dean's 'coercive' category in its focus on incentivising paid employment over benefit receipt through 'sticks' such as sanctions, increased conditionality and reductions to benefits as well as 'carrots' such as tax credits which are very much 'in work' transfers. Bonoli, like Dean, also includes a human capital category which focuses on education and training. 'Employment assistance' as a form of activation strives to facilitate re-entry into employment through more direct measures such as job placements, subsidies and assistance with job search, as is therefore similar to Dean's 'insertion' category. Finally, the 'occupation' category includes activation approaches which aim to prevent unemployed individuals becoming 'inactive' through public sector job creation and training programmes which do not necessarily have an employment-focus. This is not dissimilar to Dean's 'job creation' cluster. These distinctions are important in conceptualising 'activation' and 'active labour market policies' as a diverse range of approaches, as opposed to a homogenous whole. This book will return to these classifications in the chapters regarding EU15 nation's active ageing policies. However, as Dean notes, this taxonomy is a 'heuristic device' and therefore these categories are ideal types and as such may not exist in their pure form. As will be explored in Chapters 4–7, it is certainly the case that EU15 nations opted for a mix of these approaches in creating policies for older individuals.

Active ageing policies which focus on the employment of older people can therefore be included as part of this broader shift towards 'active' labour market policies. When initially conceived, active labour market policies (ALMPs) focused on young unemployed individuals. However, as time has progressed, new groups of formerly 'deserving' welfare recipients have increasingly felt the carrots and sticks of ALMPs. New activation policies focusing on lone parents, disabled individuals and those over 50 years of age have emerged, all emphasising the merits of, and duty to,

Table 2.1 Dean (2007) and Bonoli's (2010) taxonomies of ALMPs

Dean		Bonoli		Examples
ALMP type	Guiding Principles	ALMP type	Guiding Principles	
Human capital	Competitiveness, equality and inclusion	Human capital investment	Promoting skills as a means to employment.	– Education – Training
Coercive/work first	Competitiveness, social order and individual responsibility	Incentive reinforcement	Positive incentives for employment; negative incentives for benefit receipt	– Sanctions – In-work benefits – Tax credits – Conditionality
Active job creation	Equality, fairness and inclusion	Occupation	Ensure unemployed are kept active	– Public sector job creation – Non-employment-related training
Insertion/'right to work'	Inclusion, social order and individual responsibility	Employment assistance	Removing barriers to employment	– Job placements – Subsidies – Assistance with job search

Sources: Adapted from Dean (2007) and Bonoli (2010).

engage in paid work. Thus for this book, the term 'active labour market policies' (ALMPs) will be employed to encompass both activation and workfare, and in turn these are policies focused on recommodifying labour.

2.3   Ageing populations, early exit and intergenerational conflict

With regard to 'active ageing' policies' rationale, they have in recent years been presented as a solution to the predicted problems caused by demographic ageing and the increased frequency of early exit. Jessop (1994, 1998, 1999, 2002) also argues that population ageing is one of the exogenous pressures prompting this shift within welfare arrangements towards 'active' labour market policies but in some quarters, these changes in the age structure of the population have been related to the idea that intergenerational conflict will arise over welfare resources. Many of the fears around the sustainability of current welfare arrangements stem from the demographic ageing of national populations across Europe. The core of the issue lies with falling fertility in addition to the trend of early retirement in a period of economic downturn. The population over 50 in the European Union is set to grow from 31.3% in 1990 to 42.2% by 2020. In addition, over the same period the numbers of people aged over 60 will increase from 19.7% to 26.7% (Pearson, 1996). Not only has life expectancy increased at birth, large numbers of individuals from the 'baby boomer' generation will be approaching 50 in an era of falling fertility. Moreover, younger people as a share of working population will decline due to rising numbers entering higher education for longer periods.[4] Finally, women's increased employment has resulted in a trade-off between child-rearing and employment (Esping-Andersen, 2000). These trends have led to what have been dubbed 'demographic time bomb' arguments concerning the viability of social security systems with shifts in dependency ratios and the predicted labour. Table 2.2 shows the mean number of years still to be lived by a man or a woman who has reached the age of 65 in EU15 nations and demonstrates a rise in all countries. This broad shift towards increased life expectancy has an impact upon dependency ratios, as demonstrated in Table 2.3.

Whereas in the 1960s, conflict arose from the opposing values of young and older adults in relation to politics and ethics, Hunt (2005) and Walker (1996a) argue conflict is now more concerned with economics in relation to the pooling and sharing of resources. These arguments around

Table 2.2　Mean life expectancy post-65 years of age[5]

Mean life expectancy post-65 years of age (Men)

	1992	1993	1994	1995	1996	1997	1998	1999	2000	2001	2002	2003
Belgium	14.6	14.4	14.8	14.8	15.0	15.2	15.2	15.4	15.5	15.8	15.8	:
Denmark	14.2	14.0	14.3	14.1	14.4	14.6	14.8	14.9	15.2	15.2	15.4	15.5
Germany	14.5	14.4	14.7	14.7	14.9	15.2	15.3	15.5	15.7	16.0	:	16.1
Ireland	13.5	13.4	13.8	13.6	13.8	14.0	14.1	14.1	14.6	15.0	15.3	15.7
Greece	15.7	15.9	16.1	16.1	16.1	16.5	16.2	16.3	16.3	16.7	16.7	:
Spain	15.8	15.8	16.0	16.0	16.1	16.2	16.1	16.1	16.6	16.8	16.8	:
Italy	15.4	15.5	15.6	15.8	16.0	16.1	16.0	16.2	16.5	:	:	:
Luxembourg	14.0	14.2	14.6	14.7	14.8	14.8	15.1	15.3	15.5	16.0	15.9	15.5
Netherlands	14.7	14.4	14.8	14.7	14.8	15.0	15.1	15.1	15.3	15.5	15.6	15.8
Austria	14.5	14.7	15.0	14.9	15.1	15.2	15.4	15.6	16.0	16.3	16.3	:
Portugal	14.3	14.1	14.6	14.6	14.5	14.8	14.8	14.9	15.3	15.6	15.6	15.6
Finland	13.9	14.0	14.6	14.5	14.6	15.0	14.9	15.1	15.5	15.7	15.8	:
Sweden	15.6	15.5	16.0	16.0	16.1	16.2	16.3	16.4	16.7	16.9	16.9	17.0
United Kingdom	14.3	14.2	14.6	14.6	14.8	15.1	15.2	15.3	15.7	15.9	16.1	:

Table 2.2 Mean life expectancy post-65 years of age – *continued*

Mean life expectancy post-65 years of age (Women)

	1992	1993	1994	1995	1996	1997	1998	1999	2000	2001	2002	2003
Belgium	18.8	18.7	19.1	19.1	19.2	19.4	19.3	19.4	19.5	19.7	19.7	:
Denmark	17.8	17.5	17.7	17.5	17.8	17.9	18.1	18.1	18.3	18.4	18.3	18.6
Germany	18.1	18.2	18.4	18.5	18.6	18.9	19.0	19.2	19.4	19.6	:	19.6
Ireland	17.2	17.0	17.3	17.3	17.3	17.5	17.6	17.5	17.8	18.2	18.6	18.9
Greece	18.0	18.3	18.4	18.4	18.6	18.9	18.5	18.7	18.3	18.6	18.7	:
Spain	19.5	19.5	19.7	19.8	19.9	20.1	20.1	20.1	20.4	20.7	20.7	:
Italy	19.2	19.3	19.4	19.6	19.8	19.8	19.9	20.1	20.4	:	:	:
Luxembourg	18.1	18.5	18.7	19.2	19.2	19.0	19.2	19.5	19.7	19.4	19.9	19.0
Netherland	19.1	18.8	19.0	19.0	19.0	19.2	19.2	19.1	19.2	19.3	19.3	19.5
Austria	18.0	18.2	18.4	18.6	18.7	18.9	19.1	19.2	19.4	19.8	19.7	:
Portugal	17.5	17.3	17.9	17.8	17.8	18.1	18.2	18.3	18.7	18.9	19.0	18.9
Finland	18.1	17.9	18.6	18.6	18.7	18.9	19.1	19.2	19.3	19.6	19.6	:
Sweden	19.2	19.1	19.7	19.6	19.7	19.9	19.9	19.9	20.0	20.1	20.0	20.3
United Kingdom	18.1	17.9	18.3	18.2	18.3	18.4	18.5	18.5	18.9	19.1	19.1	:

Source: Eurostat, 2007a.

Table 2.3   Dependency ratios in EU15 nations

	1995	2000	2005	2010	2015	2020	2030	2035	2040	2045	2050
EU15	23.0	24.3	25.9	27.5	30.1	32.8	41.2	46.3	50.0	52.0	53.2
Belgium	23.8	25.5	26.3	26.4	29.1	32.2	41.3	45.1	47.2	47.8	48.1
Denmark	22.7	22.2	22.6	24.8	28.7	31.2	37.1	40.4	42.1	42.0	40.0
Germany	22.5	23.9	27.8	31.0	32.0	35.1	46.0	52.6	54.6	54.9	55.8
Greece	22.2	24.2	26.8	28.0	30.3	32.5	39.1	44.3	49.8	55.2	58.8
Spain	22.3	24.5	24.5	25.4	27.7	30.0	38.9	45.9	54.3	63.2	67.5
France	23.0	24.6	25.3	25.9	29.5	33.2	40.7	44.1	46.9	47.2	47.9
Ireland	17.8	16.8	16.5	17.5	19.9	22.5	28.3	31.6	35.9	40.9	45.3
Italy	24.0	26.8	29.4	31.3	34.3	36.6	45.2	52.4	59.8	64.6	66.0
Luxembourg	20.6	21.4	21.2	21.6	22.8	24.7	31.5	35.1	36.7	36.6	36.1
Netherlands	19.3	20.0	20.7	22.2	26.0	29.0	36.7	40.3	41.6	40.2	38.6
Austria	22.5	22.9	23.6	26.3	28.1	30.3	40.8	47.1	50.4	51.5	53.2
Portugal	21.9	23.7	25.2	26.5	28.8	31.5	39.0	43.4	48.9	54.7	58.1
Finland	21.1	22.2	23.7	25.4	31.6	37.0	45.0	47.0	46.1	46.1	46.7
Sweden	27.4	26.9	26.4	28.0	32.0	34.4	38.5	40.6	41.5	41.2	40.9
United Kingdom	24.3	23.9	24.4	25.1	28.1	30.3	37.4	41.4	43.8	44.2	45.3

Source: Eurostat, 2007b.[6]

the ageing of populations and the intergenerational conflict this will engender have also been utilised by a number of supranational organisations, such as the World Bank and the EU, to necessitate reform of existing welfare arrangements. The World Bank (1994: 4) argues that the growing numbers of young poor will create intergenerational conflict as "old retirees (some of them rich)... are getting public pensions and younger workers (some of them poor)... are paying high taxes to finance these benefits and may never recoup their contributions". In addition, it is argued that pension arrangements are inequitable as individuals from upper income groups live longer, enter employment later and therefore recoup more over their lifetimes. As a result, pension arrangements do not redistribute from lifetime-rich to lifetime-poor and consequently the "gain in real income by one generation at the expense of a permanent loss to another generation is the intergenerational transfer" (ibid: 325). Individuals from certain income groups will therefore make negative transfers, which the World Bank argues is a threat to intergenerational solidarity. This rhetoric is becoming increasingly inflammatory, with older individuals no longer portrayed as needy, but greedy, garnering less sympathy from younger cohorts who may grow resentful of contributions to fund their lifestyle. This argument however undermines contributions made over individuals' working lives, and the aforementioned social contributions that are still made post-retirement in terms of unpaid care and voluntary work.

The EU too stresses that the contract between generations is under threat due to the numeric imbalance between age groups, with calls for the "[r]ecalibration of aims at the promotion of intergenerational justice and intragenerational equality" (Amitsis et al., 2003: 12). Therefore society has changed and so should the intergenerational contract: "[d]ealing with these changes will require the contribution of all those involved: new forms of solidarity must be developed between the generations, based on mutual support and the transfer of skills and experience" (Commission of the European Communities, 2005: 6). Indeed, it is contended that nations run the risk of sustaining "an unbearable weight on public expenditure or to unfair burden sharing between the generations" (European Commission, 2000: 38).

The 2003 Committee of the Regions report argues strain on welfare states can be broken down into three components: 'green' pressure from those under 19, the working age population aged 20 to 59 who are inactive (students, disabled people and retirees) and 'grey' pressure from those aged 60-plus. Care for the 'green' generation is viewed as an investment which can be recouped via their future involvement in the labour market whereas "the elderly will play no future role in production... That they have spent the bulk of their lives creating wealth is usually ignored... in social policy terms,... financial support must be strictly rationed and controlled" (Phillipson, 1982: 17). The World Bank (1994) also argued existing welfare systems were outmoded and hard to reform in line with societal and demographic change.

In addition to the ageing of Europe's populations, another cause for concern arises from the participation rates of older individuals in the labour market. Macnicol (2004) attributes the renewed focus on inter-generational relations and the potential for conflict to the declining activity rates for those aged 50 to 64, despite the fact that this trend began over one hundred years ago. At the macro-level, early exit and retirement policies do not have wholly positive effects in that they create non-wage labour costs which hamper competitiveness, and if unemployment remains persistent, their existence results in increased social security expenditure (Ebbinghaus, 2001). Therefore aside from demographic ageing, an additional pressure on welfare arrangements is the increasingly early exit from the labour market, facilitated by state and occupational early retirement and exit schemes. Though Casey (1998) outlines a variety of determinants of early exit including health, the availability of alternative activities, mandatory retirement and earn-ings' rules, and social norms, the main overarching determinant of exit is the economic situation of the nations, which establishes the degree to which exit routes are available. As a result of the economic crises of the 1970s and 1980s, Casey argues, early exit has become entrenched in many welfare states due to the establishment of early retirement schemes.

2.4   Intergenerational solidarity and the reserve army of labour

However, the aforementioned gloomy predictions regarding the future of the welfare state in the face of demographic ageing and the impact on intergenerational solidarity have their critics. For Estes (1999a), ageing policy is a battleground with opposing forces of capitalism and demo-cracy; she argues the former utilises the discourses of the 'demographic timebomb' to erode the welfare rights secured by the latter. Indeed, she argues population ageing and the associated negative effects are now defined as 'facts', which obscures their political nature: "[t]he socially constructed 'problem', and the remedies invoked on the policy level, are related, first, to the capacity of powerful and strategically located interests and classes to define 'the problem' and to press their views into public consciousness and, second, to the objective facts of the situation. Note the order of influence: power and class, first; facts, second" (Estes, 1999a: 135). Increased longevity and an ageing population should instead be viewed as an achievement in terms of medical advances and improved standards of living (Titmuss, 1955).

For Walker (1996b), the rationale behind policy retrenchment in areas such as pensions has little to do with intergenerational conflict and argues demographic timebomb arguments are manufactured so as to provide the case for reform in line with 'blame avoidance' (c.f. Pierson, 2004). International agencies, Walker argues, are complicit in the fabricated crisis and during the 1980s, "concern about population ageing has been artificiality amplified as an economic-demographic imperative intended primarily to legitimize policies aimed at creating a new social contract between age cohorts and, more generally, restructuring the welfare state" (Walker, 1996b: 20). As a result, older people in receipt of services had to be grateful for what was on offer and younger cohorts were encouraged to provide for their own futures.

Walker (1996b) therefore argues that to some extent the potential for intergenerational conflict has been overplayed. Phillipson (1996) asserts there is no evidence of a workers' revolt against older dependents and thus any intergenerational conflict is constructed for political ends. Retrenchment of services for older people can be justified if it is on the back of public demands, and thus if governments argue they are taking preventative steps to minimise the potential for conflict by making minor modifications now, their actions can be justified and viewed as prescient. Indeed, Phillipson (1996: 217) argues "[t]he subsequent break with the generational contract was not the result of conflict between generations. Instead it arose from changes (and conflict) at the level of the state as to how this population change should be assimilated". Walker (1996b: 16) concurs, "[s]tripped of all its euphemisms the newly emerging contract between age cohorts in some western countries consists of cuts in social security for both current and future pensioners and reductions in rights of access to welfare".[7] Neo-liberal nations altered their social contracts the most, utilising arguments around 'intergenerational conflict' to provide the rationale for the reduction of benefits and services that had previously catered for those in later life. Interestingly, 'intergenerational justice' was absent from the rhetoric, perhaps because spending was more or less evenly distributed across age cohorts. Indeed, Walker argues as productivity and standards of living rise, older people should receive increases to their pension in line with real incomes, as opposed to inflation, thereby allowing them to share in the prosperity of their nation.

For Phillipson (1996), inequities between generations in terms of their current contributions need not necessarily lead to conflict. Phillipson argues intergenerational solidarity should not be seen in terms of a struggle for scarce resources but more in terms of shared needs that occur over the lifecourse. For Hills (1996), the welfare state distributes income across the individual's life cycle as opposed to redistribution from one individual to another. However, what an individual has contributed over their lifecourse is not repaid to them in old age in exact quantities as inflation means to give people exactly what they paid in would consign them to poverty and social exclusion.

Kohli (2005) argues the concerns around intergenerational conflict mask the more persistent intragenerational conflict that results from income disparities *within* cohorts. Indeed, "the discourse of generational equity overstates the extent and inevitability of such conflicts, and sharpens them at the expense of conflicts along the more traditional cleavages of class" (Kohli, 2005: 525). In so doing, Kohli argues the discourses around intergenerational conflict may in fact represent an attack from the right on the foundation of the welfare state.

These arguments would suggest reform of policies for older people is part of a more complex interplay of political motivations as opposed to a reaction to demographic ageing and its impact on welfare arrangements and solidarity. An alternative to the arguments which suggest the drive for active ageing is a product of necessity due to an ageing population and the early exit trend, or even that it is part of a broader shift towards the recommodification of labour, is a Marxist perspective which suggests older workers make up part of a 'reserve army of labour' (c.f. Marx, 1973), vilified and glorified at times of boom and slump respectively. Phillipson (2005) argues the ascendancy of capital gains over the needs of older people has meant that the priorities of the market guide policies. This, he argues, is exemplified through the cycling of older workers into and out of the labour market as is required by the economy. Therefore opportunities for early exit and retirement did not necessarily represent benevolence towards older workers, deemed to have earned a period of leisure; instead they were part of a particular welfare solution for the problems of a particular time; the role of older people as 'passive' or 'active' members of a society is in flux.

Thus Drury (1993) argues that the labour shortages in Europe in the 1960s meant older workers were seen as dependable and as they were living longer, healthier lives, their participation in the labour market was deemed acceptable. Pearson (1996) argues the economic crisis and high unemployment of the 1970s and 1980s fashioned the early exit trend which ultimately ejected and permanently excluded older workers from the labour market. In addition, states also introduced these policies to in part reduce the numbers of older individuals on the unemployment register (Mirkin, 1987). As such, Taylor and Walker (1996) argue early exit should not be seen as a form of retirement but as unemployment and by characterising it as the former, governments legitimised the focus of employment measures on those who are young and excluded from the labour market. Indeed, Ebbinghaus (2001) dubs early exit policy as a 'panacea' applied to the problem of surplus labour whilst Rein and Klaus (1993) argue early exit serves the interests of employers as opposed to the wider economy, allowing the restructuring of firms, as opposed to reducing youth unemployment (see also Samorodov for ILO, 1999). Although intended to be an 'active' labour market policy through the creation of vacancies for younger individuals, "it became a very costly passive labor market policy resulting in insufficient use, if not active disregard, of human capital and expertise" (Ebbinghaus, 2006: 203). The 'lump of labour' fallacy (Samuelson, 1980) demonstrates that the vacancies left by older workers will not automatically be filled by the young unemployed. Thus the decommodifying welfare arrangement of early exit and retirement were no longer the appropriate policy response, and thus a more 'active' approach was necessitated.

Early exit is now therefore problematised when the economy requires labour and despite calls for its end, barriers to employment remain as the belief that older workers are neither in need of nor capable of employment are internalised and institutionalised. The early exit trend has proved stubborn with Phillipson (2004) arguing institutional ageism is the result of a 'cultural lag' in that early exit was introduced at a time when there was a surplus of labour and attitudes have yet to shift to the new reality of the ageing workforce and labour shortages. Indeed, "[p]resent policies in respect of older workers still generally consider them as a labour reserve rather than as active labour market participants. The widely used 'early retirement' approach is virtually a 'hidden non-employment' approach" (Samorodov for ILO, 1999: 32).

Linked to the changes in the treatment of older workers by social policy *vis-à-vis* exit and participation are notions of rights, responsibilities and desert. In periods of labour surplus when early exit is presented positively, "'[o]bligation' meant not working" (Macnicol, 2004: 301). In times of labour shortage, as is that which many nations currently face, older individuals have the responsibility to remain in the labour market and undertake whatever measures may be necessary to achieve this (for example, training and education). At the same time, a challenge facing the drive towards active ageing is that early exit is now a social norm for many: "[i]n most developed countries the idea of a 'right to retire' is deeply embedded in the popular consciousness. Changing such views is acknowledged to be a difficult project requiring carrots (incentives) and sticks (punitive policies and practices)" (Mann, 2007: 285). Thus the active ageing agenda faces significant challenges in terms of entrenched early exit and retirement norms and policies.

Thus the shift towards recommodifying or 'active' unemployment policies from decommodifying 'passive' arrangements for older people can either be seen as a response to the changing context of welfare (as Sections 2.2 and 2.3 would suggest) or part of broader, cyclical movements which respond to the needs of the labour market. The empirical chapters of this volume will explore the extent to which this shift is apparent in EU15 nations and whether all groups within the 'older' age category are equally subject to the move to recommodification. It therefore will provide an opportunity to empirically address and refine both the recommodification and reserve army of labour literatures.

2.5   Active ageing evaluated

This chapter has explored both the resurgence of the active ageing concept and some possible explanations for its revival. In addition to the more structural explanations which have focused on the necessity of active ageing to secure the sustainability of welfare arrangements in the face of an ageing population and early exit trends, there is a significant body of literature which has argued welfare arrangements have shifted in focus from decommodifying to recommodifying labour; the narrower, economistic active ageing agenda can be seen as part of this refocusing of social policies towards promoting labour market participation. From the 'reserve army of labour' literature, older individuals are only depicted as burdensome and unproductive in periods of low unemployment; in times of labour shortage, their experience is deemed beneficial and they are portrayed as dependable. Thus the relationship between the life cycle and the labour market is not static and is the result of socially constructed notions of productivity and dependency. The pendulum has swung towards participation and 'active ageing', thereby altering the socially constructed roles for older individuals.

The rationale behind the move to 'active ageing' is one issue this chapter has aimed to explore yet the main focus of this book is not *why* the EU focuses on encouraging employment in later life, but *whether* this ethos is visible in the reforms of retirement and employment policies for older workers in EU15 countries. Just because active ageing is now on the EU's agenda, it does not necessarily follow that all nations will adopt similar policy mixes, or that all older individuals will experience this policy shift equally. The subsequent chapter will elaborate on the EU's active ageing approach which will provide a guide for the empirical chapters to follow.

The questioning of traditional notions of ageing and passivity could potentially be positive, as if "ageing in later life is perceived to be associated with declining value, increasing vulnerability and powerlessness, then the older person may appear a natural victim of modern society" (Bytheway, 1995: 5). Pearson (1996: 21) concurs that though people are living longer, healthier lives, their psychological wellbeing is undermined by society's conception of them as "an increasingly spent force and economic cost, rather than an experienced asset". Perhaps as the population ages, definitions of the 'grey' segment and what is expected of them should be altered. When initially conceived, retirement came at the end of a long career to be followed by at most a decade before death. The provision of a state pension at a set age provided a normative framework for retirement and therefore the notion of a 'normal' exit age is a fallacy in that the institution of retirement is relatively a new social construction (Marshall and Taylor, 2005). With increased longevity, retirement does not accompany a period of decline as it once did and therefore "is increasingly useless as a definition of old age" (Walker and Naegele, 1999: 2). Yet due to rigid state pension ages, functional capacity is not the determining criteria for the length of an individual's working life; instead their chronological age is the deciding factor. Retirement therefore became the third stage in the lifecourse, after education and employment and before dependent old age. As longevity increased, this period of retirement pre-dependent old age expanded. The term *'troisième age'* was coined in France to apply to this period of relative good health and social participation (Guillemard and Rein, 1993).

Guillemard (2001) argues that the neat divisions between education, employment and retirement are coming unravelled with the change in economy from Fordism and the increased flexibility of working lives. In addition, traditional notions of retirement have been complicated by early exit – both voluntary and involuntary – as well as the end of 'a job for life' (Hunt, 2005). Individuals do not necessarily travel through the lifecourse in a linear fashion – education may be returned to, sabbaticals taken – yet at the same time welfare arrangements retain their adherence to chronological age as markers of the distinct life phases. As a result, Guillemard argues, the welfare state is out of step with the new flexible lifecourse.

However, the question then becomes what is the antithesis of 'passivity' for older individuals? If 'activity' is inclusion in the labour market, should it be at any cost for any job, regardless of the skill level or emotional and financial reward it provides? For Papadopoulos (2005), the shift towards activation stresses paid work as the best provision of social protection, ignoring the paradox presented by increasingly flexible and insecure working practices. 'Security' and work are synonymous and welfare is now a helping hand which the poor are duty-bound to embrace, marking a return to the 'blame the victim' argumentation of unemployment. The state no longer provides the conditions for full employment in line with Keynesian economics, but now ensures *full employability* through the creation of flexibilised, market-ready citizens. The social right to social protection has been eroded and is now contingent upon participation in the market.

The work-focused approach to active ageing can be seen as inequitable in two main ways (based on White's critique of workfare policies (2000, 2004)). First, as the only work deemed valid is within the labour market, it ignores care work and volunteering, both of which have a significant social value and are indeed key to the broader active ageing agenda exemplified by the WHO. Second, to utilise a fairness-based work ethic (i.e. all members of society should work so no one is idle while others toil), similar employment requirements are not applied to, for example, those who have inherited large sums of money and are not engaged in paid labour. Thus "workfare forces the asset-poor to work, while leaving the asset-rich (or at least the inheritance-rich) free to ignore any such duty" (White, 2004: 278). Therefore, given the unequal access to employment and early exit opportunities, individuals are not at equally subject to work-focused active ageing policies.

The type of work individuals are encouraged to engage in so as to be 'active' is an important consideration. Attas and De-Shalit (2004) arguments regarding workfare can be applied to employment-focused active ageing in that it ignores the individual cost in terms of the quality of employment. Pearson (1996) argues a dual labour market has emerged contrasting core employment characterised by security, good conditions and terms of pay and a periphery, typified by insecurity and lower wages. Indeed, Hunt (2005) argues the shift to a service economy has created two types of employment *within* this sector.[8] The first type, in the 'primary service sector', is scarcer, offers a relatively long career with security and high incomes. In the 'secondary service sector', the jobs are low-skilled, provide minimal benefits to employees and present limited security. These jobs have a high risk of alienation and job dissatisfaction. These secondary service sector jobs are often mistaken for white-collar employment but in actuality their wages are more similar to the manual work they replaced. As Figure 1.1 demonstrates using the ILO's International Standards of Occupations, the wages of secondary service sector jobs such as service, shop and market sales employees are lower than manual labour including employment such as craft and related trades work; plant and machine operation and assembly; and elementary occupations. Therefore, for those employed in the secondary service sector, employment is unrewarding, low paid and has long hours, making it hard to escape through leisure time, which is important for forging identities. If older workers are disproportionately consigned to employment within the secondary service sector due to the aforementioned ageist attitudes of employers, this form of 'activity' will do little in terms of financial or psychological wellbeing.

Figure 2.1   Average Annual Income in Euros by Occupation (2007)

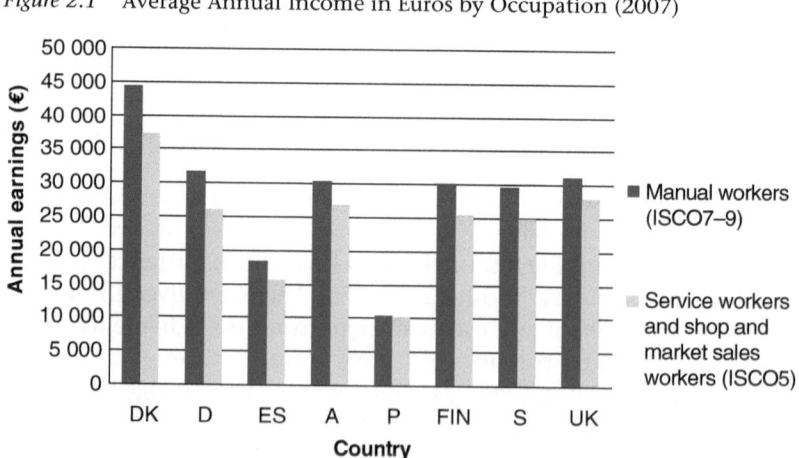

Source: Eurostat, 2010.
Note: For lists of the International Standards of Occupations, please refer to:
http://www2.warwick.ac.uk/fac/soc/ier/research/isco88/english/groups/
DK – Denmark; D – Germany; ES – Spain; A – Austria; P – Portugal; FIN – Finland;
S – Sweden; UK – United Kingdom

At the same time however, it would be unjust to suggest that early exit or 'decommodification' is always a positive experience. Though for some early exit from the labour market represents a positive choice, the degree of autonomy over participation is not equally distributed; typically the more choice a person has over their participation in the labour market, the more positive the experience of retirement will be. In terms of occupational schemes, the more advantaged a position one has in the labour market, the more advantaged a position one typically has with regard to autonomy over labour market exit. Those with 'nest eggs' to leave for or offered a generous voluntary redundancy packages are in a better position than those who find they are the first to go when their employer makes job cuts. However it can also be the case that older unemployed people "once their 'golden handshakes' have been exhausted, experience many years of poverty, with no chance of re-entering the labour market, except in the most menial of jobs" (Phillipson, 1982: 3).

Older people relying on the state for income, Hunt (2005: 55) argues, "become, in essence, second class citizens. Not being part of the consumer world is therefore a form of social exclusion". In the context of early exit and extended retirement, "the 'third age' is linked today with increasing uncertainty, both concerning the degree to which old age will remain socially and financially secure, and whether people have a chance to succeed in living an old age which is meaningful" (Evers and Wolf, 1999: 42). Employment is important in Hunt's account as the financial rewards of paid work provide life chances and mark the ascension into adult life. It provides an important facet of individuals' identities and "our sense of self-esteem and personal well-being is wrapped up in the work that we undertake" (Hunt, 2005: 147). Indeed, those older age individuals who are expelled from the labour market with limited opportunity for re-entry may find the experience detrimental. However, in a post-modern society, Hunt argues what we consume is more integral to one's identity than what we actually do. Thus for Hunt, work, leisure and consumption are three key themes and older individuals may be unduly disadvantaged in these areas. The end of employment heralds the end of social ties and ushers in a period of potential isolation. Employment is a means of participating and contributing to society as well as a source of social identity; its cessation may lead to alienation. Thus the active ageing agenda can be seen as a response to the growing need for individuals to work longer to guarantee a secure and comfortable old age.

Thus perhaps what is the most important consideration is the degree of choice individuals have regarding their commodification, decommodification or recommodification. Indeed,

> [a]gency is a key element of the life course... but agency has to be conceived as 'agency within structure'... Individuals are embedded in structures and institutions. Institutions can be and often are sources of agency by providing continuity and coherence of biographical orientations. Welfare programs, in particular, provide competencies, resources, opportunities, and individual rights that empower individuals outside and inside of the market and the family (Leisering, 2003: 219).

This book therefore also addresses the degree of choice individuals have regarding their labour and examines whether certain groups are disproportionately constrained within the context of the shift towards recommodifying welfare policies. Thus this book will not only make general statements regarding particular nations' policy approaches whilst neglecting the differences at the micro-level in terms of the options available to different groups of older individuals. It would be oversimplistic to state that nations are moving towards the active ageing agenda without interrogating what that means at the micro-level: are some groups expected to age more actively than others? Do other groups retain the ability to withdraw from the labour market? If, as the arguments presented before related to intergenerational conflict, active ageing is a means to promote intergenerational solidarity, does this come at the price of greater *intragenerational* inequality?

The following chapter will narrow the focus to examine the EU's perspective on 'active ageing' before outlining the schemas which will be used to address EU15 nations' policy mixes and the impact they have on older individuals' autonomy with regard to labour market participation.

3
The EU's Active Ageing Agenda

This chapter will begin by exploring the EU's 'active ageing' discourse and proposed policy approach. As the preceding chapter outlined, active ageing policies can embody more than the governance of retirement to include areas such as leisure, voluntary work, health and wellbeing. For this book however, as the focus is on EU15 nations, it is important to examine the EU-level policy discourse on active ageing. In aiming to address the extent to which EU15 countries have converged towards the EU's active ageing policy agenda, this book acknowledges that it does not necessarily follow that all nations will adopt the same sets of policies, or that the EU has caused any policy change. This chapter therefore includes arguments to suggest why differences in national approaches may persist, both in terms of the policy mixes they present, and in terms of how these welfare arrangements interact with different groups within the older age cohort. Finally, this chapter will outline the schema which will be used to address EU15 nations' progress towards the EU's active ageing approach.

3.1   The EU's active ageing discourse and agenda

The pervading argument throughout European Union documents is that changes brought about by globalisation have made the creation of a New Social Policy Agenda imperative. The new agenda embodies a "strong focus on high social standards and a strong safety net... [to]... form the cornerstones of the policy admixture which will drive the European social model forward over the next few years" (European Communities, 2000: 13). At the same time, this new agenda cannot be realised without a substantial shift towards full employment (see the Lisbon strategy, European Commission, 2000; European Council, 2000). The participation of older workers is seen as integral for medium- and long-term economic growth and the accomplishment of the 70% total employment target (Council of the European Union, 2004). In addition, societal ageing is argued within EU documents to lead to a decline in the working population and thus increase pressure on pensions systems and health care (Commission of the European Communities, 1999: 6). The shift in national demographics resulting from falling fertility, increased longevity and the 'baby boomer' generation entering the third age have necessitated that social protection systems must become more attuned to these demands through a process of modernisation (Commission of the European Communities, 2005). The contract between generations is also argued in many EU level documents to be under threat from the numeric imbalance between the generations, with Amitsis et al. (2003: 12) in a report for the EU calling for the "[r]ecalibration aims at the promotion of intergenerational justice and intragenerational equality".

As part of the broader full employment agenda, 'active ageing' first appeared in EU documents in 1999 (Commission of the European Communities, 1999: 6) with regard to reinforcing the employability of older workers and adapting employment rules to the ageing population. This document emphasised retirement as a constrained decision: "[o]ver the working lifetime, their risk of marginalisation and eventual exclusion from the labour market grows. In the end, older workers often find that early retirement is the only choice left to them" (ibid: 10). Yet at the same time, it suggests access to early exit routes, such as extended unemployment benefit durations, should be limited to "reduce the demand for and the access to early retirement schemes" (ibid: 13). Since 1999, active ageing has become a feature of EU/EC discourse yet definitions of measures that can be included under the umbrella of active ageing vary within policy documents. However, its core principles can be regarded thus: a focus on lifelong learning; policies to encourage working longer, retiring later; being active after retirement; and finally engaging in capacity enhancing and health sustaining activities (Commission of the European Communities, 2002). The responsibility for the delivery of this approach includes three levels: employers, social security systems and civil society, as demonstrated below:

- At the level of employment and social security systems
  - decrease the incentives to leave working life early and to reduce strongly early retirement
  - promote later retirement through pension reforms
  - develop the possibility for retired people to work (including through part-time or temporary jobs)

- At the company level, in particular through the involvement of the social partners
  - promote the implementation of lifelong learning for older workers
  - improve working conditions
  - modernise the organisation of work in particular to better meet the needs of older workers while effectively using their expertise, including through shifting types of jobs (e.g. from a management post to an advisory post or coaching) or by taking up employment in a dependent company
- At the level of society
  - To increase the employment rate of older workers will require society to think differently about the potential contribution of older workers; this will imply a tremendous change of mentalities
  - Promote a shift in public opinion through for instance advertisement campaigns (like in Nordic countries) (EC, 2004: 80).

Thus the first of these – the policies introduced at the level of employment and social security systems – will be the focus of this book. Though in some EU documents, 'active ageing' includes a wider range of measures and services, encompassing lifelong learning; policies to encourage working longer, retiring later; being active after retirement; and engaging in capacity enhancing and health sustaining activities (Commission of the European Communities, 2002), the end is ultimately to extend the working lives of those approaching retirement age. As Avramov and Maskova (2003) argue, recent years have seen the definition of 'activity' focus firmly on labour market participation, especially in the policy discourses employed by the EU.

This focus is highlighted by a number of EU targets and summits. The 2001 Laeken Summit in Brussels created 11 common objectives to secure adequate and sustainable pensions in Europe, and included increasing the employment rate for older individuals. Concurrently, the Ageing Working Group under the Economic Policy Committee reported in its findings on public pension expenditure from 2000–2050 that extending working lives was main policy approach that would combat rising pension costs. It was clear therefore that the early exit trend would have to be reversed and the strategy of retiring older workers to reduce unemployment would no longer be viable (Von Nordheim, 2004). This realisation and the EU's active ageing approach are reflected in two targets. At the 2001 Stockholm European Council, it was announced that the aim was to ensure half of those aged 55 to 64 are in employment by 2010. The following year in Barcelona, the target for an increase in the

Table 3.1   Employment rates for 55–64 year olds, EU15 nations

	'95	'96	'97	'98	'99	'00	'01	'02	'03	'04	'05	'06	'07	'08	'09	'10
Austria	29.7	29.1	28.3	28.4	29.7	28.8	28.9	29.1	30.3	28.8	31.8	35.5	38.6	41.0	41.1	42.4
Belgium	22.9	21.9	22.1	22.9	24.6	26.3	25.1	26.6	28.1	30.0	31.8	32.0	34.4	34.5	35.3	37.3
Denmark	49.8	49.1	51.7	52.0	54.5	55.7	58.0	57.9	60.2	60.3	59.5	60.7	58.6	57.3	57.5	57.6
Finland	34.4	35.4	35.6	36.2	39.0	41.6	45.7	47.8	49.6	50.9	52.7	54.5	55.0	56.5	55.5	56.2
France	29.6	29.4	29.0	28.3	28.8	29.9	31.9	34.7	37.0	37.8	38.5	38.1	38.2	38.2	38.8	39.7
Germany	37.7	37.9	38.1	37.7	37.8	37.6	37.9	38.9	39.9	41.8	45.4	48.4	51.5	53.8	56.2	57.7
Greece	41.0	41.2	41.0	39.0	39.3	39.0	38.2	39.2	41.3	39.4	41.6	42.3	42.4	42.8	42.2	42.3
Iceland	39.2	39.7	40.4	41.7	43.7	45.3	46.8	48.0	49.0	49.5	51.6	53.1	53.8	53.7	51.0	50.0
Italy	28.4	28.6	27.9	27.7	27.6	27.7	28.0	28.9	30.3	30.5	31.4	32.5	33.8	34.4	35.7	36.6
Luxembourg	23.7	22.9	23.9	25.1	26.4	26.7	25.6	28.1	30.3	30.4	31.7	33.2	32.0	34.1	38.2	39.6
Netherlands	28.9	30.5	32.0	33.9	36.4	38.2	39.6	42.3	44.3	45.2	46.1	47.7	50.9	53.0	55.1	53.7
Portugal	46.0	47.3	48.5	49.6	50.1	50.7	50.2	51.4	51.6	50.3	50.5	50.1	50.9	50.8	49.7	49.2
Spain	32.3	33.2	34.1	35.1	35.0	37.0	39.2	39.6	40.7	41.3	43.1	44.1	44.6	45.6	44.1	43.6
Sweden	62.0	63.4	62.6	63.0	63.9	64.9	66.7	68.0	68.6	69.1	69.4	69.6	70.0	70.1	70.0	70.5
United Kingdom	47.5	47.7	48.3	49.0	49.6	50.7	52.2	53.4	55.4	56.2	56.8	57.3	57.4	58.0	57.5	57.1
EU15	36.0	36.3	36.4	36.6	37.1	37.8	38.8	40.2	41.7	42.6	44.2	45.3	46.5	47.4	47.9	48.4

Source: Eurostat, 2011.

average retirement age by five years before 2010 was set. These goals and active ageing by extension are also seen as a means to achieve the Lisbon strategy's goal of 70% employment. As Table 3.1 demonstrates, some EU15 nations had already exceeded the Stockholm target in 2001.

To summarise the EU's guidance for active ageing social policy measures, Table 3.2 has been compiled from prescriptions in the technical annex of the 2006 EC report *Adequate and Sustainable Pensions*.

3.2   Policy divergence: The macro-level

Explanations have been presented for the recent resurgence in the concept of active ageing in the previous chapter, and the EU's interpretation has been outlined above. However, though the EU has embraced active ageing and produced a number of recommended policy changes, it does not necessarily follow that all EU15 nations will adopt similar approaches. These nations are similar with respect to their membership

Table 3.2   EU's active ageing approach

Policy area	Examples
Implementing strict rules for eligibility for old-age pensions	– Increase eligibility ages; – increase contribution thresholds; – both in conjunction.
Rewarding deferred retirement and discouraging early retirement	– "Motivating people to make their own choices about when they retire can be considered even more efficient, than restricting the exit from labour markets by statutory retirement ages or financial disincentives" (European Commission, 2006: 34); – Higher accrual rates for deferred retirement either for working beyond state pension age or contribution threshold; – Abolition/reduction of early retirement schemes; – Stricter criteria for early pensions – disability or long contribution records; – Barriers to longer working removed – age discrimination legislation.
Providing flexibility in retirement	– "Eroding the cliff-edge between working and retirement (which has had negative impacts for generations of workers) is an important element of increasing employment rates for older workers" (European Commission, 2006: 36); – part-time work/reduction of working hours.

Source: Based on recommendations in 2006 EC report *Adequate and Sustainable Pensions*.

of the European Union and the creation of "common institutions to which they delegate some of their sovereignty so that decisions on specific matters of joint interest can be made democratically at European level".[1] They were also member states when the Lisbon strategy was implemented, including its goal of 70% employment as well as the Stockholm and Barcelona targets which relate specifically to older individuals.

In addition, the nations face similar demographic changes (though to different degrees, as demonstrated in Tables 2.2 and 2.3) and produce National Action Plans on the topic of employment in the hope of creating cross-national dialogue in the realm of social policy. The EU also creates directives on certain areas, such as anti-discrimination legislation, which nations are then given a fixed amount of time to adopt. These directives tend to prescribe certain policy ends but leave the means to the discretion of the nation-states.

However, social policy formation is incredibly complex with many of causal influences and as a result, to assume the EU's active ageing model has *caused* national policy changes would result in 'crude empiricism' (Doyal and Harris, 1986 in Spicker, 2008). Therefore in terms of convergence, this volume avoids overstating the role of the EU in social policy. Indeed, though this book uses the model of active ageing as proposed by the EU as a benchmark against which to address national approaches, it does not seek to explore what has *caused* these changes and is therefore influenced by the new institutionalist literature (see Pierson, 2004) which argues that the focus on policy convergence underplays national divergence and the importance of policy legacies. Their perspective is anti-functionalist, arguing an institution does not only survive as long as it retains utility. Indeed, institutions often survive due to 'path-dependency' in that once the die has been cast, the game must continue from there. Thus Pierson's (2004) argument is used to elucidate institutional inertia and the notion of 'increasing returns' explains why nations or institutions may get 'locked into' certain policy paths as once they have started down a particular route, the costs of deviation may be more than continuation, thus making dramatic reform difficult. Pierson argues the likelihood of change is impeded by this process and can only be achieved by three means: cumulative causes, threshold effects and causal chains. Indeed, the direction of change is constrained by previous actions, thus explaining the nation-specific responses to similar crises. Nations are therefore circumscribed within 'menus' of institutional change.[2]

3.3 Policy divergence: The micro-level

Just as this book will avoid overstating the role of the EU in altering nations' active ageing approaches, it too will not overstate changes at the macro-level and ignore differences at the micro-level insofar as the different policies available to groups within the category of 'older age' will be explored. The political economy of ageing literature is a key influence here, emphasising how different groups experience ageing in diverse ways and how the state in turn has a role in mediating these processes.

This literature utilises both the conflict and critical theory to focus on how economic and political structures determine the allocation of resources, and how this in turn shapes individual experiences of ageing (Bengtson et al., 2005; Estes, 1999a). Thus "constructions of aging and the social policies that result not only <u>reflect</u>, but also <u>reproduce</u> existing social class, gender, and racial and ethnic disparities among the old. That is, social policy on aging presently does little to alter or disturb the relations of power or the distribution of economic and other resources in the society" (original emphasis, Estes, 1999b: 136). As a result, the political economy of ageing approach focuses on two key aspects: how different social classes, genders and ethnicities experience ageing and how the state is involved in the construction of old age. Though ageing is a biological fact, many of the social facets are constructions and are experienced differently by the various sub-groups that make up a particular cohort (Walker and Phillipson, 1986). In addition, "[s]ocial policies are shown to constrain citizens' preferences, cement patterns of social inequalities and slow social change" (Meyer and Pfau-Effinger, 2006: 69). Thus social policy defines who is deserving and undeserving which perpetuates existing social divisions.

For example, 'decommodification' through early exit or retirement in older age is not equally distributed across all sub-groups within the age category. In terms of state-provided early retirement policies and extended benefit durations or job search exemptions which provide *de facto* exit, when welfare provision is deferred to all citizens, it is a social right; when the individual behaviours of citizens are included it becomes a question of desert.[3] In terms of the latter, welfare arrangements create subjects to be governed, as was seen with early Poor Laws such as the British Poor Law of 1834 and the Dutch Armenwet of 1854 which divided the needy into 'deserving' and 'undeserving' and distributed relief accordingly. Fischer (2003: 67) argues "by separating populations into the deserving and undeserving groups, politicians are able to legitimize the bestowal of beneficial regulations or subsidies on the former and punishment or neglect on the latter". The provision of early exit schemes with eligibility criteria such as contribution thresholds or employment in certain industries similarly divide older workers into deserving and undeserving. Phillipson (1982) argues it is those older individuals in lower socio-economic stratum that are at the most risk of being classified as 'undeserving' and marginalised.

Individual biographies affect the ability to access welfare policies. For example, in terms of gender, the experiences of older women are socially constructed, characterised by choices that were highly constrained within and by three spheres: the state, the market and the family. Indeed, "[g]ender is a crucial organizing principle in the economic and power relations of the social institutions of the family, the state, the market, shaping the experience of old age and ageing and the distribution of resources to older men and women across the life-course" (Estes, 2005: 552). Women's experience of the labour market is often entwined with their family life (Carr and Sheridan, 2001; Estes, 1991; Heinz, 2001). Gender norms traditionally emphasised the female role in caregiving within the family, which was reinforced in many nations by welfare arrangements (Bettio and Plantenga, 2004; Millar, 1999; Pfau-Effinger, 1999). Inequalities related to gender are carried through into old age as due to women's interrupted or part-time careers, they are often dependent upon their husbands in terms of state and occupational pensions to provide an adequate standard of living in old age (Meyer, 1998). In addition, the effects of sexism in employment can persist into retirement as "the retired executive or doctor, who will more often be a man, will have accrued throughout his life considerable assets as a result of house ownership, investment of savings and occupational pension rights. He will be less dependent on the state" (Bond, 1986: 48). Thus older women do not have equal freedom of choice over their exit and participation in labour market, which has a knock-on effect upon wellbeing in later life.

However, it has been argued there is therefore variation internationally in terms of the effect of care on women's employment. Bettio and Plantenga (2004) argue national care strategies have implications in terms of social and economic outcomes for women. In their analysis of European countries they found the cluster containing Greece, Italy, Spain, Portugal and Ireland, with its lack of provision for care, inhibited female participation in the labour market which is particularly pronounced among those with low-skills whose employment would not allow for the purchase of care services. Indeed, a study by Harkness and Waldfogel (1999) demonstrated that in comparison with childless women and men, women aged 25–44 were less likely to be engaged in employment in Germany, France, Sweden, Finland and the UK, yet there was significant variation between the nations. In the case of the UK, the difference between the two groups of women was greatest, with 76% of those without children in full-time employment, compared to 26% of those with children. On the other hand, in the case of Sweden, this difference was less marked at 75% for childless women compared to 61% for those with children (see also Ginn, 2004). In addition, though states may encourage women to engage in paid labour, as long as their employment prospects are not equal to those of men, their pensions too will remain unequal (Meyer, 1998). Women also make up the majority of part-time workers, who receive in general lower wages (Frericks and Maier, 2008).

A further important caveat is the importance of cohort membership as Bengtson et al. (2005) highlight in determining the roles women occupy *vis-à-vis* employment and care; younger generations may be less inclined and less expected to engage solely in unpaid care. As a result, there is considerable variation amongst older women in terms of their choice with regard to retirement. Indeed, those who entered the labour market in the 1960s (and were therefore born in the 1940s), work was still very much divided along gender lines and women often exited permanently with the arrival of children. These women had unequal access to vocational education and as a result their employment income was not sufficient to provide independence. However, those who were born in the 1960s and entered the labour market in the 1980s faced a very different picture in that opportunities for education and work had improved (Yeandle, 2001; see also Meyer and Pfau-Effinger (2006) for a methodological application of this argument). Though women from the 'baby boomer' generation are better educated and healthier than previous cohorts, their degree of choice regarding labour market participation is still impeded by three constraints including financial resources, care responsibilities and functional abilities (Arber, 2006).

Other filters through which social policies are experienced include class and ethnicity. In terms of the former, pensions are key in providing income in old age and the benefits of occupational schemes are contingent on employment history, creating what Titmuss (1955) termed 'two nations in retirement', with those with resources more likely to be younger and middle class whilst those who are female, working class and single are found to be lacking financially. As the experience of ageing is bound to the individual's labour market history, disadvantage and discrimination faced by minority ethnic groups during their working lives are carried through into old age. Nazroo (2006) however argues literature which suggests race and age present a 'double jeopardy' is over-deterministic.

3.4   Exploring active ageing and policy change in EU15 nations

With this in mind and using the policy prescriptions outlined in Table 3.2, the schema in Table 3.3 (constructed using various EU and EC documents and influenced by De Vroom's (2004a) work) guided both the data collection and analysis. As this book focuses specifically on EU15, it was logical to limit the focus to the active ageing as envisaged by this supra-national organisation. To allow for the presentation of what was a great deal of data in a coherent way, the policies have been divided into two distinct categories within which there are additional sub-categories:

- Pension policies:
    - Pension principles and state pension age/s;
    - Early retirement;
    - Deferred retirement;
- Labour market policies:
    - Early exit (including unemployment benefit extension and supplements, and job search exemption);
    - Active labour market policies (ALMPs);
    - Part-time work/pension arrangements.

Table 3.3 demonstrates the direction the reform of these policies areas should take to exemplify the EU's active ageing approach.

Early exit and early retirement are not the same thing, though in a practical sense they both represent withdrawal from the labour market. Guillemard and Rein (1993) make the distinction between early retirement and early exit: the former is when an individual enters an old-age pension scheme; the latter is when an individual has stopped working but has not entered an old age pension scheme. Thus policies such as extended benefit durations and supplements as well as job search exemptions represent *de facto* early exit routes; though they do not explicitly allow for the permanent exit from paid employment, in effect older individuals are able to remain detached from the labour market

Table 3.3   Configuration of the EU's active ageing agenda

Pension policies	The **reform** of state pension ages.	– Increase eligibility ages (European Commission, 2006). – Increase contribution thresholds (European Commission, 2006). – Flexible retirement options (European Commission, 2006).
	The **retrenchment** of early retirement routes.	– Abolition/reduction of early retirement schemes (European Commission, 2006); – Stricter criteria for early pensions – disability or long contribution records (European Commission, 2006).
	The **expansion** of policies to encourage deferment.	– Higher accrual rates for deferred retirement either for working beyond state pension age or contribution threshold (European Commission, 2006).
Labour market policies	The **retrenchment** of *de facto* early exit, i.e. unemployment benefit extensions/supplements and job search exemptions.	– Closing off early exit routes (European Commission, 2006).
	The **expansion** of active labour market policies for older individuals.	– Barriers to longer working removed – age discrimination legislation (Council of the European Union, 2000; European Commission, 2006). – Part-time work/reduction of working hours (European Commission, 2006). – Training (Committee of the Regions, 2003). – Incentives to remain in employment (European Commission, 2006).

for an extended period. There are also *de jure* early withdrawal schemes in the form of early retirement pensions. Those schemes that enabled individuals to exit or retire early for the purpose of this book are considered to be policies that allowed for the decommodification of labour; thus reforms and retrenchment working in the opposite direction, i.e. restricting access to these routes through adjustment to eligibility criteria, making them less attractive through the application of penalties as well as their closure represent the *de facto* recommodification of labour through the removal of alternatives to paid employment. In addition, policies designed to reintegrate and retain older individuals in the labour market such as tailored ALMPs and incentives to defer pension receipt are also defined in this book as policies aimed at recommodifying the labour of older individuals (however, these are *de jure* recommodification policies).

3.5   Progress towards active ageing: The macro-level

From the literature outlined in Chapter 2 and above, two questions emerge: first, can a shift away from early exit and retirement policies towards ALMPs for older individuals be observed in EU15 nations, thereby embodying the EU's 'active ageing' approach? Second, as the political economy of ageing literature suggests, do individuals within the older age group experience the active ageing policy differently? To put it another way, are some older individuals expected to age more actively than others? This book will address both of these questions,

Table 3.4   EU15 nations and their employment rates for 55–64 age group in 2002

	Country	Participation rate (%) 2002
	EU15 average	40.2
Group II: 'Particularly worrisome' with less than 35% of older people in work	Belgium	26.6
	Luxembourg	28.1
	Italy	28.9
	Austria	29.1
	France	34.7
Group III: 'The in-between group', close to the EU average (40.2%)	Germany	38.9
	Greece	39.2
	Spain	39.6
	Netherlands	42.3
	Finland	47.8
Group I: Close to or above the Stockholm target	Ireland	48.0
	Portugal	51.4
	United Kingdom	53.4
	Denmark	57.9
	Sweden	68.0

Source: Data from Eurostat (2011); categorisation from Council of the European Union (2004).

first by exploring the policy approaches to older individuals more generally in the EU15 nations from the mid-1990s (and thus prior to the emergence of the active ageing agenda) until 2010 (the deadline of the Barcelona and Stockholm targets) and second by focusing on the degree of choice different sub-groups within the 55–64 age cohort had with regard to labour market participation or exit.

In 2004, the EU divided its member states in terms of their position in relation to the Stockholm target of achieving 50% of 55–64 year olds in employment by 2010 as of 2002, i.e. one year after the launch of the goal (Council of the European Union, 2004). The nations were divided accordingly (see Table 3.4):

- Group I – seems to be close to, or even above, the Stockholm target, among them, Sweden, Denmark, the UK,... Ireland,... and Portugal.
- ... Group II – is particularly worrisome, with... Belgium,... Luxembourg, Italy, Austria and France, where less than 35% of older people were at work.
- 'in-between' group – group III – consisting of Germany,... Finland,... Spain, the Netherlands, Greece..., is close to EU average (Commission of the European Communities, 2004: 6–7).

Since the early 2000s, much has changed with regard to this categorisation, as Table 3.5 demonstrates. Of those countries in Group II, the 'worrisome' nations for whom employment rates for the 55–64 age group were below 35%, progress has been made by all, most notably Austria, Luxembourg and Belgium. Nations in Group III whose employment rate for the 55–64 year olds had been under 50% but close to the EU average (38.8%) on the other hand made some mixed progress towards the target. Germany, Netherlands and Finland had by 2010 surpassed 50% whilst Spain and Greece remain below, having made limited increases of 4.4 and 4.1% respectively. Those nations close to or beyond the Stockholm target have also made different levels of progress, indeed Portugal and Denmark have actually seen a decline in participation rates by –1% and –0.4% each.

As a result of these changes, a new categorisation would seem appropriate, based both on EU15 nations' position with regard to the Stockholm target in 2010 but also in terms of their progress towards over the nine year period from when the goal was set in 2001.[4] Indeed, some of the nations in the original Group II with employment rates for the 55–64 year olds lower than 35% would effectively have had further to travel to reach the Stockholm target. Across EU15 nations, the

Table 3.5   EU15 nations and their employment rates for 55–64 age group 2001 and 2010 compared

		2001	2010	Increase/ decrease
	EU15	38.8	48.4	+9.6
Group II: 'Particularly worrisome' with less than 35% of older people in work	Belgium	25.1	37.3	+12.2
	Luxembourg	25.6	39.6	+14.0
	Italy	28.0	36.6	+8.6
	Austria	28.9	42.4	+13.5
	France	31.9	39.7	+7.8
Group III: 'The in-between group', close to the EU average (40.2%)	Germany	37.9	57.7	+19.8
	Greece	38.2	42.3	+4.1
	Spain	39.2	43.6	+4.4
	Netherlands	39.6	53.7	+14.1
	Finland	45.7	56.2	+10.5
Group I: Close to or above the Stockholm target	Ireland	46.8	50.0	+3.2
	Portugal	50.2	49.2	−1.0
	United Kingdom	52.2	57.1	+4.9
	Denmark	58.0	57.6	−0.4
	Sweden	66.7	70.5	+3.8

Source: Data from Eurostat (2011); categorisation from Council of the European Union (2004).

employment rate for the 55–64 age category increased by 9.6%; some nations still below the Stockholm target have made greater increases including Belgium, Luxembourg and Austria; some close to the EU average in 2002 (38.8%) and therefore in Group II have far surpassed it (Germany, the Netherlands and Finland) while others have failed to keep up as it rose to 48.4% in 2010 (Greece, Spain). Thus what this volume suggests is the following as a way of organising and presenting a great deal of empirical data on national policy pictures (see Table 3.6):

- *Group I – The Vanguards:* Nations that were in 2001 above or close to the Stockholm target and have since maintained or strengthened this position, including Sweden, Denmark, the UK, Ireland and Portugal.
- *Group II – Surpassing Stockholm:* Nations that since 2001 have moved beyond the Stockholm target including the Netherlands, Finland and Germany.
- *Group III – Below Stockholm but approaching fast*: Nations that have not yet met the Stockholm target, but have made progress beyond

Table 3.6 EU15 nations classified according to the progress towards the Stockholm target 2001–2010

	Country	Employment rate (%)		Increase/ decrease
		2001	2010	
	EU15	38.8	48.4	+9.6
Group I: Above the Stockholm target in 2001	Sweden	66.7	70.5	+3.8
	United Kingdom	52.2	57.1	+4.9
	Denmark	58.0	57.6	−0.4
	Ireland	46.8	50.0	+3.2
	Portugal	50.2	49.2	−1.0
Group II: Nations that since 2001 have moved beyond the Stockholm target	Germany	37.9	57.7	+19.8
	Finland	45.7	56.2	+10.5
	Netherlands	39.6	53.7	+14.1
Group III: Nations that have not yet met the Stockholm target, but have made progress beyond the average percentage change for EU15 nations between 2001–2010 (9.6%)	Austria	28.9	42.4	+13.5
	Luxembourg	25.6	39.6	+14.0
	Belgium	25.1	37.3	+12.2
Group IV: Nations that have not yet met the Stockholm target and whose progress was less than the average for EU15 nations (9.6%)	Spain	39.2	43.6	+4.4
	Greece	38.2	42.3	+4.1
	France	31.9	39.7	+7.8
	Italy	28.0	36.6	+8.6

Source: Data from Eurostat (2011); classification author's own.

the average percentage change for EU15 nations between 2001–2010 (+9.6%), including Belgium, Austria and Luxembourg.

– *Group IV – The Laggards*: Nations that have not yet met the Stockholm target and whose progress was less than the average for EU15 nations (+9.6%), including France, Greece, Italy and Spain.

3.6 The micro-level: Exploring the impact of reform on individual autonomy

In addition, in order to examine the impact of policy changes at the micro-level, this book will employ 'model biographies'. Influenced by approaches utilised by Meyer et al.'s (2007) edited volume on state and private pensions and Bradshaw et al.'s (1993) model family approach, this aspect will address the impact of employment history and age on the policy choices individuals have pre- and post the launch of the EU's active ageing agenda.[5] However, whereas the works of Bradshaw et al. (1993), Meyer et al. (2007) and Meyer and Pfau-Effinger (2006) focused on outcomes of welfare arrangements, this book addresses the outputs in terms of the choice policies provide individuals in relation to their labour market exit or participation (decommodification or recommodification). The data will not elaborate on the net gains and losses of particular policies and as aforementioned, policies that provide exit with only a very low level of income could not be considered positive. However, the presence of such a policy increases the choice an individual has, albeit perhaps in a symbolic as opposed to a practical way. Yet the inclusion of quantitative measures of losses and gains that could be incurred by the various policies would not automatically provide an insight into the losses and gains at the individual level. To assume an individual would take the most financially expedient policy option would rely too readily on a rational choice model. Indeed, individuals may prefer to utilise an early exit route, though it represents a loss of income, as it would provide the freedom to supply care or engage in fulfilling voluntary work. Conversely, the presence of generous early exit routes may not mean an individual would automatically leave paid work, which may provide self-fulfilment and social networks. In addition, in weighing up employment versus early exit in financial terms, an assumption would have to be made about the income from employment, which would limit the model biographies analysis to polarised occupations or would result in an extremely large data matrix.

Exit from the labour market does not represent a homogenous experience and this is because the older age cohort itself is made up of diverse groups who will experience ageing, employment and retirement differently. In turn, state policies will mediate their experience and will not interact with all sub-groups equally. Thus the aim is to address the various eligibility criteria and conditions applied to groups within the older age cohort to explore the differential policy treatment of these sub-groups and avoid sweeping generalisations about the recommodification of older individuals' labour as a whole. In addition, as the literature review outlined, the level of labour market participation required of older individuals in terms of whether 'responsibility' referred to engagement or exit is also in flux, and policy plays a role in prescribing levels of acceptable activity and passivity in old age. Therefore the inclusion of model biography data will address which groups within the older age cohort are deemed by policy to be deserving of exit or decommodification and who are in need of activating through recommodifying welfare arrangements. This begs the question, if (as the policy discourse at national and supra-national levels implies) active ageing is a universal good, both for the individual and for society, why do some individuals retain the ability to become 'passive' and decommodified?

Different contribution records were included in the model biographies used by this book to address the impact of labour market participation on the choices older individuals have regarding their de- and recommodification through pension and labour market policies. An individual with a disjointed employment history could represent a woman with caring responsibilities, yet equally it could represent a man with a recurrent health problem, or an individual from a minority ethnic group subject to repeated racial discrimination in the workplace and consigned to a series of temporary jobs. In terms of the inclusion of two ages, this attempted to establish whether within the older age cohort there was a difference with regard to decommodification and recommodification options, i.e. were nations more likely to have and retain decommodification options for older individuals and whether younger individuals were subject to recommodification policies. In addition, in some nations, state pensions were (and remained) unharmonised for men and women, thus utilising gender as a principle upon which decommodification and recommodification was bestowed. The model biographies utilised were as follows:

- the 50-plus:
  - *Laurent:* Aged 55 with 35 years of employment contributions.
  - *Jean:* Aged 55 with a disjointed employment history.
- the 60-plus:
  - *Sasha:* Aged 63 years with 35 years of employment contributions.
  - *Jude:* Aged 63 with a disjointed employment history.

What will become apparent in the empirical sections of this book is that certain characteristics influence an individual's interactions with the state and state policies; age, gender and occupation all effect the level of labour market participation an individual is expected to engage in (decommodification (exit) or recommodification (entry)) in the EU15 nations.

This chapter has explored the EU's approach to active ageing, which as a concept began to appear in EU documents from 1999 onwards. Both the EU discourse around active ageing and the targets it has set regarding longer working lives and retaining older workers in the labour market have emphasised the centrality of paid work to remaining 'active' in older age. This book seeks to examine the extent to which EU15 nations have moved towards the EU's active ageing agenda both at the national and micro-levels. In other words, it seeks to address whether EU15 nations are encouraging the labour market participation of older people, and whether there are differences in the policy options available to groups within this age cohort. The following four chapters look at each of the four groups of EU15 nations outlined in Table 3.6, exploring the national policy pictures before comparing the approaches adopted and the policy options available at the micro-level.

4
Group I: The Vanguards – Consolidating Their Position Beyond Stockholm

As aforementioned, Sweden, Denmark, the UK and Portugal had all surpassed the Stockholm target of 50% employment for older individuals and Ireland was deemed by the EU to be close at 48% in 2002 (Council of the European Union, 2004). This chapter will explore the policy changes implemented in these nations from the mid-1990s until 2010, considering the policy context prior to the launch of the Stockholm and Barcelona targets as well as the subsequent reforms. It will then examine the impact of these changes to policies for work and retirement for the sub-groups within the older age category.

4.1   Sweden

In 2001 when the Stockholm target was established, Sweden had an employment rate of 66.7% for 55–64 year olds – the highest of all EU15 nations. By 2010, it had risen to 70.5% – beyond the EU Lisbon target of full employment for the population in general, and is still therefore the highest of EU15 nations (Eurostat, 2011). The following sections will elaborate on the pension and unemployment policies applicable to older people available from the mid-1990s, prior to the establishment of the Stockholm target, until the deadline of 2010.

Pension policies
Pension principles and the state pension age
Sweden represented a vanguard with regard to pensions, having introduced the first universal public system in Europe. The initial pension system comprised of two elements: a national flat rate basic pension based on residence in Sweden and the national income-related supplementary pension scheme (*tilläggspension*, ATP). The latter was financed

by employers' contributions. The pension age was reduced in 1976 to 65 from 67 years of age.

The effective retirement age for men decreased from 67 in 1965 to 62 in 1998 whilst it rose for women from 55 to 61 over the same time period (Committee of the Regions, 2003). As a result of this decline in male participation, a ruling in June 1998 meant that as of 1999 a new retirement system was introduced, with its first payments coming into effect in 2003. Under the old system, pensions were flat rate and for those who had lived in Sweden for between 30 and 40 years. There was however, an earnings-related pay-as-you-go (PAYG) element. From 2003 onwards, pension amounts would correspond to the people's earnings since the age of 16, thus encouraging individuals to maximise their earnings over their lifecourse. In addition, the scheme shifted from a defined benefit to a defined contribution pension. There were contribution-based elements: a PAYG earnings-related old age pension (*inkomstgrunded ålderspension*) with fixed contributions of 16% of the pensionable earnings and a fully-funded premium reserve system (*premierservsystem*) which operated on an insurance principle. In addition, there was still a basic citizenship pension – the guarantee pension (*garantipension*) – which was for those with no or a low earnings-related pension (MISSOC, 2005). Individuals born before 1937 would receive their pension in accordance with the previous system; those born between 1938 and 1953 would receive their pension in accordance with a combination of the two systems, with the proportions depending on the year of their birth, and those born after 1953 would be subject to the new system (OECD, 2000). The new system will be fully in place by 2015 (Committee of the Regions, 2003).

Early retirement

In 1970, the disability insurance scheme was reformed to include special eligibility rules for those aged between 63–67 years of age. The medical assessment was more lenient for this age group and they were not required to engage in retraining if their health was poor. Two years later, labour market considerations were taken into account in that individuals over the age of 63 would be eligible if they had exhausted their entitlement to unemployment benefits, regardless of their health status. In 1974, the age threshold was lowered to 60 to reflect the reduction in the state pension age to 65 which would come into effect in 1976. The 1990s saw a change in policy direction with the condition regarding the exhaustion of unemployment benefits repealed in 1991 and the special conditions for older people removed entirely by 1997 (Jönsson et al., 2011; Palme and Svensson, 1999).[1]

In addition, Sweden endorsed flexible retirement with pensions payable fully or partially at 61 with the option of moving in and out of retirement. Individuals could exit between the ages of 60 and 64 with a reduction of 0.5% for every month shy of the state pension age. As part of the 2003 reforms, the lower age threshold for the flexible early retirement scheme had been raised to 61 (Committee of the Regions, 2003). In spite of the presence of this scheme, Sweden nonetheless had comparatively high employment rates for older individuals; indeed, even as early as their entry to the EU in 1995, the participation rate for those over 55 was 62% compared to the EU15 average of 36% (Eurostat, 2010).

Deferred retirement

Throughout the 1990s and 2000s, in order to extend working lives, individuals who deferred pension claims and received consent from their employer could continue working until the age of 70 and receive an increase of 0.7% per month over the age of 65 to a maximum of 42% (OECD, 1997, 2005m).

Labour market policies

Early exit

As of 1974, in Sweden older workers who were made redundant at 58.3 years old could remain on unemployment benefits until they reached the age of 60, when they could then access the disability insurance scheme until the state pension age. These '58.3' pensions were closed off by 1992 due to the cost (Mandin, 2004). In 1997 an additional early exit option was created for the long-term unemployed who were 60 years older or more who had been unemployed for two years. These individuals could remain on unemployment benefits until they reached the state pension age or older which made it possible for those with a history of being out of work for two years or more to get a special compensation up to the normal retirement age of 65. It was however, only available for the second half of 1997 and yet despite its short lifespan, 18,956 individuals opted for this early exit compensation (Sjögren, 2006; Wadensjö and Sjögren, 2000; Wadensjö, 2002).

In terms of the unemployment benefits available, as with many of the EU15 nations, there were two schemes in place. In the 1990s, both the assistance and insurance schemes had extended durations for older individuals (for unemployment insurance, those under 55 could access the benefit for 300 days whereas those over 55 could receive it to 450 days; unemployment assistance could be accessed for 150 days by those under 55, 300 days for those aged 55–59 and 450 days for those over 60). The unemployment benefit system was reformed by 2000 to replace the unemployment assistance with the basic allowance (*grund-försäkring*) and the unemployment insurance with an optional earn-ings-related benefit (*inkomstbortfallsförsäkring*). However, there was still the option of an extended benefit duration though the age threshold had been raised to only include those over the age of 57. By the mid-2000s, this option had however been abolished.

Active labour market policies

Though Sweden did retain an option to retire early, at the same time, their commitment to promoting employment does in some part explain the high participation rates for older workers. Due to Sweden's 'work society' tradition, everyone is considered to have the right to work. As a result, Swedish law works to promote this and employment policy is more 'active' than 'passive'. In 1975, a Bill was introduced to provide individuals aged 45 to 65 with a longer notice period for dismissal (Mandin, 2004). The Employment Protection Act in 1982 created the right to work until the age of 67. This was however undermined by the collective agreements of social partners who could terminate the contracts of their employees at 65 years of age.

In terms of active labour market policies, in 1996 temporary public sector jobs were created for those who were over 54 and long-term unemployed under the OTA (*Offentliga Tillfälliga Anställningar*) scheme (Wiklund, 1998; Sweden, 1999). However, after its initial introduction, no new placements were created and thus the numbers enrolled had tailed off to 124 individuals by 2001 (Wadensjö, 2002). Many older individuals registered at the job centre only received a basic education of nine years in compulsory schooling (Wiklund, 1998). Measures to provide a 'knowledge lift' (*Kunskapslyftet*) for those with limited formal education over the age of 25 by "improving basic skills and knowledge, the project aims to increase self-confidence among less well educated people, in order to build a more solid base for their future learning in work settings, thereby helping them to be more flexible in a constantly changing labour market" (Tikkanen, 1998b: 112) ran from 1997 to 2002.

To promote the re-employment of older individuals, from August 2000, for those who had been unemployed for 24 months and over the age of 57, any potential employer received a 75% tax reduction for the first two years of their employment. This scheme, the Special Employ-ment Allowance, had a threshold so that the amount could not exceed the ceiling of SEK (Swedish Krona) 525 per day, which roughly corresponded to half of the average wage for full-time workers. For all other age groups, to benefit from this scheme they had to have been unemployed for 48 months and receive a tax deduction of 75% for the first year, reduced to 50% for the following year (OECD, 2005m, 2006; Committee of the Regions, 2003). In addition, to encourage those who had retired early to re-enter the labour market, from 2000 people could return to work for a maximum of three years without affecting their pension amount (Regeringskansliet, 2002).

Part-time work/pension arrangements

In 1976, the partial pension (*delpension*) was introduced for individuals moving to part-time work from the age of 60 (Anderson and Immergut, 2007). Individuals between the ages of 60 and 64 had to reduce their working time to between 17 and 35 hours per week and the loss of income was reimbursed at 55% (MISSOC, 1996). The individual was required to have worked for at least ten years since the age of 45 (Wadensjö, 2002). In 1998 a reform was introduced so that no further partial pensions could be bestowed after 2001 (Wadensjö, 2002; OECD, 2005m; Committee of the Regions, 2003). However, the pension reform in 2003 altered this to allow individuals to claim $\frac{1}{4}$, $\frac{1}{2}$ or $\frac{3}{4}$ of their pension and curtail their working hours accordingly, increasing the flexibility of exit from the labour market (Committee of the Regions, 2003).

4.2 Denmark

Denmark too exceeded the Stockholm target when it was set in 2001 but by 2010, had experienced a slight drop in the employment rates for the 55–64 age group from 58% to 57.6%. Unlike Sweden which had made significant gains in the employment rate for this age group, in Denmark it has remained fairly stable for nearly 20 years as the graph below demonstrates, only dipping below 50% between 1995 and 1996 and surpassing 60% in 2003–4 and 2006 (Eurostat, 2011).

Pension policies

Pension principles and the state pension age

The Danish pension system comprises of three pillars: the first includes a flat-rate PAYG national pension (*folkepension*) and a labour market supplementary pension (*arbejdsmarkedets tillægspension*, ATP), the second is occupational schemes and the third is individual savings (Belloni et al., 2006). Denmark's National Pension (*Folkepension*) was viewed as

Figure 4.1   Employment Rate for 55–64 year olds in Denmark 1992–2010

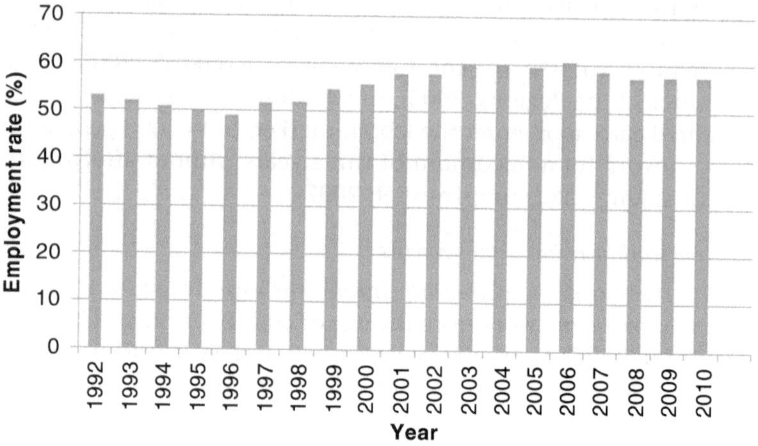

Source: Data from Eurostat, 2011.

a right of all citizens, in addition to the Supplementary Pension (*arbejds-markedets tillægspension*, ATP) which was compulsory for all individuals working more than nine hours per week. Those on benefits, including early pensions could however opt to contribute to the latter scheme. The National Pension provided a flat-rate benefit whilst the Supplementary Pension was linked to contribution years. Comparatively, these pensions had high age thresholds at 67 years of age for both genders. Around one-third of all pensioners in 1998 had supplementary pensions which will continue to increase due to an agreement in 1991 between Trade Unions and employers to provide them for the majority of employed individuals. This, Walker (1993) argues, disadvantages women who are more likely to have disjointed work histories than men and therefore cannot fully exploit these occupational pensions. Denmark has taken steps to redress the imbalance by providing pension credits for those who exited the labour market to provide care or were on maternity leave.

In July 1999, the state pension age was reduced from 67 to 65 for those who at this point were not over the age of 60. Thus, the first effects of this policy were seen in 2004 (Bertelsmann-Stiftung Foundation, 1999). This, Jensen (2004) argues, will reduce public expenditure as the level of pension is low in comparison to early retirement or disablement benefits. In addition, "the reduction in pensionable age is not intended to encourage people to withdraw from the labour market at the age of 65. Rather the contrary is true. In order to encourage pensioners to remain in the labour market after the age of 65, the deduction from income (wage work, etc.) in the old age pension is to be reduced from 60 per cent to 30 per cent" (Jensen, 2004: 43). However, reform undertaken in 2006 will raise this pension age back to 67 by 2022 (Bredgaard and Tros, 2006).

Early retirement

In Denmark, there were two main early retirement routes. The first, the early social pension/anticipatory pension (*førtidspension*), allowed individuals to exit early following a medical statement certifying a loss of employability.[2] This scheme was introduced in 1984 and merged the invalidity and widowers' pensions. This pension could be received by those aged from 50 to state pension age (67 in 1995) on the grounds of health (incapacity of at least 50%) or social (unemployment, poor training or limited income prospects) reasons. For individuals under 60, there was the basic sum (*grundbeløbet*), a pension supplement (*pensionstillægget*) and early retirement sum (*førtidsbeløbet*) (Tikkanen, 1998b; Jensen, 2004; MISSOC, 2001).

Second, the voluntary early retirement pension/VERP (*efterløn*) allowed the individual to choose to retire between the ages of 60 and 67 (Jensen, 2004). This pension was introduced in 1979 and the rate was similar to that of the unemployment benefits. When originally initiated in the late 1970s, the early retirement scheme served two purposes. First, it offered a route out of the labour market in a period of high unemployment. Second, the scheme redistributed retirement opportunities in a more equitable manner. It was argued that all individuals contributed to the conventional pension system but benefited differentially in that those in unskilled, blue-collar employment would be more likely to die before or soon after the state pension age. Thus the early pension allowed those in arduous work to exit early and reap the rewards offered by retirement (Jensen, 2004). This pension had a contribution requirement of 20 of the previous 25 years and the benefit was distributed in two steps: in the first stage of two and a half years it was equivalent to the unemployment benefits, followed by a reduced second phase at 80% of unemployment benefits. The individual was permitted to work up to 200 hours per year (Hansen, 2002; Larson and Pedersen, 2005).

In addition, the transitional allowance or 'very' early retirement scheme also known as the transitional income-related pension (*overgangsydelse*) was introduced in 1992 for those aged 55 to 59 who were long-term unemployed yet had been a member of an unemployment insurance scheme for 20 years over their working lives. The benefit amount corresponded to 80% of the unemployment benefit; the same as for phase two of the voluntary early retirement pension/VERP (*efter-løn*) (Hansen, 2002; Jensen, 2004). Individuals enrolled on this pro-gramme received a transitional allowance until the state pension age of 67; these individuals could not enter the early retirement scheme that commenced at 60 years of age (Drury, 1993). This rule was altered in 1994 to include those from the age of 50 upwards (OECD, 2005c).

Between 1996 and 2000, a series of changes were introduced regarding the early exit routes and labour market opportunities for older individuals. With regard to the transitional income-related pension (*overgangsydelse*), in 1994 8,250 people were enrolled, rising rapidly to 23,294 in 1995 (Venge, 1998). As a result of the high take-up, this policy was barred to new entrants in 1996 but those already enrolled could continue until they reached the age of 60, thus allowing the scheme to be entirely phased out by 2006 (OECD, 2006). It was the hope that by taking this action, a further 26,000 persons would enter the labour market (Ministry of Economic Affairs and Ministry of Labour, 2002: 14). In addition, older workers were encouraged to engage in activation measures (see below).

The VERP (*efterløn*) underwent reform in 1998 to strengthen the link between work and pensions. To incentivise longer working lives, indi-viduals who remained in employment until the age of 62 received tax deductions and the full pension amount until the official state pension age of 67 (Bertelsmann-Stiftung Foundation, 2002b). From the 1st of July 1998, the anticipatory pension scheme (*førtidspension*) was reformed to become more stringent by stating pensions could not be awarded until activation measures aimed at improving an individual's working capacity had been undertaken.

Prior to the 1998 election, the Danish Prime Minister Poul Nyrup Rasmussen of the Social Democratic party pledged to maintain the early exit routes that were available. However, upon re-election, these policies were altered. Jensen (2004) argues the Trade Unions of Denmark tradi-tionally saw early retirement as their domain but conceded in times of socio-economic change, reform was necessary; the major changes of 1999 were deemed essential given the shifting socio-economic context. In addi-tion to the reforms of unemployment schemes of 1999 (see below), in this year on the 27th of April the Danish Parliament proposed the amend-ment of early retirement schemes. It was suggested that remaining in the labour market longer should receive greater financial encouragement, as well as the option to combine part-time work and pensions (Ministry of Economic Affairs and Ministry of Labour, 1999). A series of changes were implemented which affected the various early retirement schemes to different degrees. The first, the VERP saw its eligibility ceiling raised to

25 years of membership to the unemployment insurance fund of the previous 30 years. In order to discourage the use of the VERP, in 1999 the pension reduction for those exiting between 60 and 62 was 10% (OECD, 2006). Additionally, a lump-sum bonus of 8,600 DKK (Danish Krone) per quarter was introduced in this year for those who continued to work between 62 and 65 years of age (OECD, 2000). New entrants had to make a special contribution to the unemployment insurance scheme in order to receive early retirement (Jensen, 2004). The benefit level was raised; when the old scheme was introduced, it corresponded to 100% of unemployment insurance benefits, falling to 82% over a number of years. The new scheme increased the rate to 91%. It is argued by the OECD (2005c) that this reform could not raise the VERP age further so this bonus was the alternative means to encourage prolonged labour market participation. Despite this reform, however, the VERP remained the primary early exit route (OECD, 2005c).

Though the early retirement certificates which bestowed the right to early retirement upon the sick (early social pension/anticipatory pension (*førtidspension*)) were not altered in 1999, a year later they was reformed to place those who would have previously been eligible to retire early in subsidised 'flexjobs' (see active labour market policies (ALMPs)). This was an extension of a scheme introduced for the labour market in general in 1994 (Jensen, 2004). This pension scheme was reformed again as of January 2003, following the changes instigated in 2000 as part of a move to a more active approach, focusing on the capabilities of the individual and utilising them in a job placement (Ministry of Economic Affairs and Ministry of Labour, 1998). The individual had to receive this benefit on a full-time basis, having been deemed incapable of part-time employment (OECD, 2006). The following parameters are considered to limit the working capacity of the individual: "vocational and practical competences, personal competences to enter into social and job relations, networks of labour market relevance, health and the individual's own job perspectives" (Ministry of Economic Affairs and Ministry of Labour, 2002: 39). These capabilities were assessed and developed to better tailor employment measures for the individual. From this year onwards, the caseworkers who authorised the award of this pension were required to provide a clear rationale in each case. Thus individuals with reduced working capacity would first and foremost be offered a subsided, 'flexjob' on a full-time basis before an early pension (OECD, 2005c; Ministry of Economic Affairs and Ministry of Labour, 2002).

By 2005, in order to access the VERP, an individual had to have been eligible for unemployment benefits and the pension they received must be equal to 91% of this state transfer. Those who postponed access to VERP for two years and continued to work for 30 hours a week received a tax-free bonus. The reforms of 1999 had made it less financially attractive to exit the labour market early, but the Liberal-Conservative Government in collaboration with the Social Democrats went further in 2006. As part of these reforms, the early retirement age will be raised to 62 from 60, and the state pension will again be increased to 67 from 65. These changes will take full effect in 2022 (Bredgaard and Tros, 2006).

Deferred retirement

In 2004, new incentives to extend working lives were introduced. Those over the age of 65 were then able to defer their pensions for up to ten years and receive an increase to the final amount of 7%. However, the individual had also to work for at least 1,500 hours per year, which was above the national average (OECD, 2005c). Spring 2008 saw the launch of Job Plan to encourage individuals to work longer. As of July 2008, the individual could defer retirement up to the age of 75 and the hours threshold had been reduced to 1,000 hours a year (Social Security Administration, 2008). In addition, individuals in employment at 64 would receive a special tax relief of up to DKK 100,000 if they had remained in full-time work from the age of 60. This scheme will run from 2010 until 2016 (Ministry of Social Welfare and Ministry of Health and Prevention, 2008).

Labour market policies

Early exit

Different rules were applied for those over the age of 50 compared to the general population with regard to the unemployment benefit durations. In general, individuals who were insured and found themselves unemployed were entitled to benefits for two periods, initially four years, and then three (MISSOC, 1999). The individual was also interviewed by a Public Employment Service officer within six months, with those felt to be at most risk of long-term unemployment targeted first. After this six month period, if the individual was still unemployed, they would be obliged to engage in an activation measure (Ministry of Economic Affairs and Ministry of Labour, 2002: 30). Individuals over the age of 50 could however, remain on benefits until they reached the age of 60 and could then access an early pension. They were also exempt from the job search requirement of benefit receipt and the right to participate in activation measures. This was known as the '50–59 rule' (Hansen, 2002). Yet for those who became unemployed over the age of 60, the unemployment benefit duration was limited to 2½ years. In addition, those 60 and over were barred from schemes that provide the start-up costs for self-employment and the 'job offer scheme'. It is as a result of these age limits that Drury (1993) argued Denmark operated age discrimination.

1999 saw the third revision of the labour market policy for unemployed individuals. With regard to insured persons, first, the obligation to participate in activation measures had been made more rigorous in terms of the rules governing the availability for work and a new threshold was set at less than 12 months of benefits receipt. The '50–59 rule' was altered in this year to only exempt those aged 55 to 59 from job search (Hansen, 2002). Under these changes, individuals would be invited for an interview before reaching three months of unemployment. With regard to older individuals, specific attention was given to those aged 50 to 59. In addition, the total maximum period of benefit receipt in general had been reduced from five to four years.

In 2001, those receiving unemployment benefits were entitled and obliged to engage in training and an individual action plan; refusal resulted in the withdrawal of benefit. Individuals aged 58 to 59 and unemployed remained exempt from activation measures (Bertelsmann-Stiftung Foundation, 2003a). Initiatives were taken in 2002 to encourage the labour market participation of older workers. First, a 'senior policy pool' was established to provide financial support to local initiatives designed to promote active ageing. Also, a senior policy consultancy scheme provided these enterprises with free advice (Ministry of Economic Affairs and Ministry of Labour, 2002).

Active labour market policies

From 1997, in an attempt to reduce early exit, the local authorities were required to first offer rehabilitation or reintegration measures, with retirement as the last resort (OECD, 1997). An individual had a duty to participate in an activation programme after two years of unemployment. The flexjob scheme was available more generally but was also offered as an option to those aged 60 and who had contributed to the unemployment insurance fund for 20 of the past 25 years, i.e. those eligible for the VERP early retirement scheme (Jensen, 2004). The senior benefits scheme was introduced in 1997 for those who exited a flexjob before the age of 65 to cover the potential gap until state pension age (Jensen, 2004). In 1997, a collective agreement between municipalities and the trade unions produced the Senior Employees Policy which allowed employees over the age of 52 to move into part-time work and maintain their pension rights (Committee of the Regions, 2003).

The individual had to be available for participation in job-training schemes in order to receive unemployment benefits as of 1998 (Ministry of Economic Affairs and Ministry of Labour, 1998). For those individuals who were over 25 and uninsured, participation in activation measures had to commence before one year of social support receipt. The municipality was obliged to formulate an action plan for the individual if they were over 25 years of age. As of July 1998, this was amended to include all age groups up to the age of 30 who then had to engage in an activation programme before being in receipt of benefits for 13 weeks (Ministry of Economic Affairs and Ministry of Labour, 1998). In 1998, an agreement was signed in Denmark between the National Association of Local Authorities and the Association of Local Government Employees' Associations to allow older workers to move to less demanding work. The trade unions opposed the proposal to reduce older workers' working hours (Taylor, 1998). In addition, the rules for income tax deductions had been simplified in 1998 so as to incentivise unsubsidised employment (Ministry of Economic Affairs and Ministry of Labour, 1999).

Between 1999 and 2001, public sector jobs were created for those who had been unemployed for $1\frac{1}{2}$ years and were over 48 years of age (Hansen, 2002). As aforementioned in the section on *Early retirement*, in January 2000 the early social pension/anticipatory pension (*førtidspension*) scheme was narrowed and individuals who would have been previously eligible were also allocated subsidised 'flexjobs' (Jensen, 2004). As of 2000, those over 60 were also entitled to and obliged to take part in activation measures before six months of unemployment (Ministry of Economic Affairs and Ministry of Labour, 1999; Hansen, 2002). Additionally, between 2000 and 2003 2.5 million DKK were available for older unemployed individuals who set up 'self-activating' support groups (Hansen, 2002).

In 2004 a new adult vocational training system was introduced to promote employability in addition to increasing the age threshold for access to state education grants from 59 to 64 (OECD, 2005c). This year also saw the state pension age effectively reduced from 67 to 65 years of age (the policy had come into existence in 1999, but only for those under 60 at the time; thus 2004 saw the first cohort aged 65 since the policies' creation). Though the transitional allowance was disbanded in 1996, individuals enrolled before this point were still present in 2004.

The individuals on the scheme could exit the labour market on 82% of unemployment benefits if aged between 55 and 59 (OECD, 2005c).

Since 2004, age discrimination legislation has been in place in Denmark. Employers were, however, still permitted to enforce mandatory retirement ages at the default of 65 or earlier, if objectively justified (OECD, 2006). More generally, Denmark aimed to make the labour market more attractive to older workers with attempts to alter the perceptions of employers to instead view this group as a valuable resource (Venge, 1998). Those who were enrolled on the unemployment register and over 50 were allowed access to the same schemes as younger unemployed individuals. An agreement was signed between the National Association of Local Authorities and Association of Local Government Employees' Associations which meant that older workers could move to less demanding work (Taylor, 1998).

As of January 2008, older individuals whose entitlement to unemployment benefits had expired had the right to a job placement in their local authority until they reached the age threshold for the voluntary early retirement pension (VERP). The individual would be employed under normal pay and conditions and the local authority received a subsidy (Ministry of Social Welfare and Ministry of Health and Prevention, 2008).

Part-time work/pension arrangements

To cater for those ineligible for a voluntary early retirement pension/ VERP (*efterløn*), the part-time (*delpension*) pension was introduced in 1987 (Jensen, 2004). This partial pension was for those aged 60 to 67 (until the state pension age was lowered in 1999, after which the upper threshold was 64) who permanently resided in Denmark and had been working for at least seven hours a week over the previous nine months and had made ten contribution years (Ebbinghaus, 2006). The reduction in working time had to be between 12 and 30 hours (OECD, 2005c).

The 1999 reforms also combined the partial pension with the VERP scheme by allowing the individual to combine work and pension receipt, providing they worked no more than 200 hours per year or 29.6 hours a week, an increase from the old part-time pension which had a limit of 27.75 hours per week (Jensen, 2004). This, it was argued, would encourage a more gradual transition into retirement (MISSOC, 2001). In addition, a new part-time pension was available for those aged 60 to 64 from 2004 and who were not eligible for the early pension. The individual had to have been in the labour market for at least 20 years and worked for nine of the past 12 months. The individual had to reduce their working time to between 12 to 30 hours per week. By 2005, the part-time pension (*delpension*) had been altered to stipulate that the individual had to have participated in the supplementary pension scheme (ATP pension) for at least ten out of the past 20 years and must have worked at least 18 out of the past 24 months in Denmark.

4.3   The United Kingdom

The UK's employment rate for the 55–64 age group was also in excess of the Stockholm target in 2001, and had risen by a further 4.9% by 2010 (Eurostat, 2011). Unlike Sweden and Denmark however the UK only exceeded 50% employment of older people as relatively recently as 2000 whilst the former two nations moved beyond this target in the early 1990s. As will be explored below, the UK is also different from Sweden and Denmark in its total lack of state-funded early exit and early retirement schemes, which has been the case since 1989 when the Job Release Scheme was closed.

Pension policies

Pension principles and state pension age

In the 1990s, the UK pension system comprised of a flat-rate basic state pension and an additional State Earnings Related Pension, accessible by women aged 60 and men aged 65. The state pension age for women was rising gradually to age 65 from 2010 to 2020 as a result of the 1995 Pensions Act; the 2011 Pensions Act has accelerated this to reach 65 by 2018 for women born after April 1953. The Pension Act of 2007 increased the state pension age for both genders to 68 by 2046, beginning in 2026 (Social Security Administration, 2010).

Early retirement

See *Early exit*.

Deferred pensions

Until April 2005, it was possible for individuals to defer state pension receipt for up to five years and resulted in an increase to the final pension amount of 7.4% per year. Following the Pensions Act of 2004, the bonus for deferral increased to 10.4% per year worked beyond the state pension age. Individuals could also take a lump sum made up of the bonus for deferral and the pension as yet not claimed (Social Security Administration, 2004).

Labour market policies

Early exit

The UK had no statutory early exit routes in place since the disbandment of the Job Release Scheme in 1989 (Committee of the Regions, 2003). The alternatives, Incapacity Benefit and Job Seekers Allowance are not specifically targeted at older individuals and are unappealing due to their low rates and conditionality (Mandin, 2004). Though Incapacity Benefit represented the main state-funded exit route in the UK, from 1995 this benefit was made less generous (OECD, 2005n). In terms of unemployment benefits, the only concession for older individuals was that the asset threshold for non-contributory unemployment benefits was higher than for the general population.

Active labour market policies

With regard to the re-skilling of older individuals, in 1993 the age threshold for the 'Training for Work' scheme was increased from 59 to 63 (Walker, 2002). This scheme provided training for long-term unemployed individuals. However, in 1995 the Training and Enterprise Councils which administered this scheme prioritised those aged 18 to 24 (Taylor, 1998). The Conservative Government of this period favoured a voluntaristic approach to age discrimination with the occasional awareness campaign, such as the 'Getting On' campaign in 1993 to educate employers about the value of older workers (Walker, 2002: 418).

In June 1999, a non-statutory *Code of Practice on Age Diversity in Employment* was published. The Code emphasised the need to abolish age restrictions and provided employers with advice as to how to best adhere to its recommendations (UK, 1999). Taylor (2005) argues the UK had the most extensive campaign of all other EU15 nations to promote the utility of older workers. After April 1999, upper age limits for training provided by Work Based Learning for Adults and Training for Work in Scotland were abolished. In addition, the career development loans had no upper age threshold and the student loan age limit was extended to those under 54 years of age, provided they aimed to return to the labour market after their spell in education (UK, 1999).

April 2000 saw the national launch of the New Deal 50 Plus (ND50+) as part of New Labour's drive for 80% employment. However, the resources allocated were a fraction of those available for younger cohorts (Loretto and White, 2004), with spending on the ND50+ only 2.1% of expenditure on all New Deal programmes (Trades Union Congress, 2005). In terms of eligibility, an individual had to be over 50 and in receipt of Income Support, Jobseeker's Allowance, Incapacity Benefit, Severe Disablement Allowance or Pension Credit for six months or more. Those in receipt of National Insurance Credits, Invalid Care Allowance or Bereavement Allowance could also be eligible. The ND50+ included access to a Personal Advisor; for those earning under £15,000 annually and working at least 16 hours a week, there was a 'fifty-plus element' to the Working Tax Credit for 52 weeks and the opportunity to use a Training Grant up to £1,500 within two years of finding employment. This programme was entirely voluntary. In addition, the state and local partnerships together set up a pilot known as the PRIME initiative to help people over the age of 50 create their own businesses in 2000 (UK, 2000).

In 2001, a number of measures were introduced in the UK to re-skill older individuals including Third Age Apprenticeships. In addition, this year saw the establishment of an Age Advisory Group of leading Social Partners to liaise with the government and establish 'good practice' with regard to active ageing (UK, 2001) and the establishment of the Age Positive campaign (Taylor, 2005). With regard to pre-existing policies, initially the ND50+ contained a weekly Employment Credit of £60 for those whose wages were below £15,000. However, this financial incentive was argued to have a 'deadweight' effect (Hirsch, 2003) and was replaced by the fifty-plus element of the Working Tax Credit in 2003. The rationale behind the fifty-plus element of the Tax Credit was that after a year in receipt, the individual would have gained higher paid employment. In addition, the unemployment benefits system was altered with the lowering of the assets threshold for the income-based jobseekers allowance to £6,000 for those over the age of 60 (for those under this age, the threshold was £3,000).

In light of concerns around ageism, in October 2006 legislation to curb discrimination in the workplace based on the European Community report *Equal Treatment in Employment and Occupation Directive* (2000) was introduced. The language of the document *Equality and Diversity: Coming of Age – Consultation on the Draft Employment Equality (Age) Regulations 2006* (DTI, 2005) was very much concerned with the needs of industry in addition to tackling discrimination to the end of social justice: "It is only fair that those who experience age discrimination should have the same opportunities as others. But equally our economy is one in which the labour market struggles to meet the appetite of business for a skilled and adaptable workforce – we cannot afford to cast on the scrap-heap some of our most experienced, skilled and valuable people on grounds of prejudice" (DTI, 2005: 3). The directive includes both older and younger workers. Mandatory retirement ages and age-specific recruitment will be prohibited unless justified. In addition, 65 remained in the 'default' retirement age and employees could then submit a request to their employer to work beyond this. Thus retiring individuals aged over 65 did not constitute discrimination, though employers could implement lower retirement ages if they could justify them. Employers had to write to their employee between 6–12 months before they intended to retire them and they in turn have the right to request to remain in employment post-65. The employer then had a 'duty to consider any requests' (DTI, 2005: 20).

As of October 2009, the ND50+ was being closed off with some areas of the UK instead operating the new Flexible New Deal for all age groups (this scheme itself was considered extremely costly and was closed in a year and replaced with the Work Programme, Department of Work and Pensions, 2011; Public and Commercial Services Union, 2011). With regard to age discrimination legalisation, the UK Coalition Government announced in July 2010 that the default retirement age would be repealed gradually from April 2011 to be closed entirely by 2011. The *Employment Equality (Repeal of Retirement Age Provisions) Regulations 2011* outlined that from 6 April 2011, employers will not be able to issue any notifications for compulsory retirement using the Default Retirement Age procedure. Older workers could only be compulsorily retired if their employer had notified them by 6th April 2011 and if their retirement fell before October 2011.

Part-time work/pension arrangements

The option of combining pension receipt and paid work was not applicable in the UK.

4.4   Ireland

The EU placed Ireland within their Group I of countries who had reached or were near to reaching the Stockholm target in 2002 (Commission of the European Communities, 2004: 8). Ireland at that time had an employment rate for older workers of 48%, which had risen to 50% by 2010. Ireland therefore had one of the lowest levels of employment for 55–64 year olds out of all Group I nations by 2010, but also in the early 1990s when it was less than 40%. This had risen to a high of 53.8% in 2007 and there had been a slight drop by 2010 (Eurostat, 2011). Ireland therefore had comparatively further to travel towards the 50% target than other Group I nations.

Pension policies

Pension principles and state pension age

The Irish pension system has comprised of three pillars since the nation became independent in 1922. The three pillars included a flat-rate state pension as well as occupational and private pensions, undertaken at the discretion of the individual. As there was no state-provided earnings-related supplement, many individuals utilised occupational schemes with private pensions increasing in popularity since 1992 (Schulze and Moran, 2007).

The first pillar of the Irish state pension system itself had three different elements (Schulze and Moran, 2007; OECD, 2005g). The Old Age Contributory Pension was available at 66 for those who had contributed to the Pay Related Social Insurance (PRSI) scheme for at least 156 contribution weeks with an annual average of at least 20 per year before the age of 56 and those with an average between 24 and 48 contribution weeks per year would receive a reduced pension amount. Second, the Retirement Pension provided a bridge to the Old Age Contributory Pension at the age of 65 for one year. It was available to a similar group as the Old Age Contributory Pension except that contributions had to have been made before the age of 55 and exit could occur at 65 years of age with a full pension. The minimum period of membership for receipt of this pension was slightly higher at 24 contribution weeks per year and as with the Old Age Contributory Pension, a reduced payment would be incurred for contributions between 24 and 48 weeks per year. Once this pension had been accessed, employment was prohibited. The third, for those over 66 who were ineligible for the former pensions, the Old Age Non-Contributory Pension was set at a flat rate and was on a means-tested basis (OECD, 2005g, 2006, MISSOC, 1994; Schulze and Moran, 2007).

Early retirement

Ireland offered no early retirement options. See below for *Early exit.*

Deferred retirement

There were no incentives for pension deferral in Ireland and work could only be combined with receipt if earnings are below a certain threshold (€38 per week).

Labour market policies

Early exit

In terms of early exit routes, the Pre-Retirement Allowance (PRETA) was established in 1990 and provided a means-tested benefit for those wishing to exit over the age of 55 and under the state pension age who had been in receipt of unemployment benefits for at least 390 days. These individuals would move from long-term Jobseekers' Allowance and would no longer be required to 'sign on' with the Social Welfare Office or Signing Centre, instead annually completing a Declaration Form (PR39). In April 2002, the PRETA scheme was reformed so as to allow those enrolled in this programme to leave to take up employment and re-enter should they subsequently become unemployed (Department of Enterprise, Trade and Employment, 2002). From July 2007, PRETA was closed to new applicants and thus individuals who would formerly have been eligible were required to 'sign on' at the employment service and were subject to the new activation measures in terms of the National Employment Action Plan (NEAP) process listed below (Grubb et al., 2009).

Active labour market policies

In 1998, Ireland introduced the Employment Equality Act which decreed that there were to be seven grounds of discriminatory action, including age. The Act covered direct and indirect discrimination in the field of employment. Indeed, "[t]he Act grants employees rights to equal remuneration and equal treatment in employment and provides for the insertion of equality clauses into all contracts of employment" (Committee of the Regions, 2003: 226). However, the Act does contain some exemptions, including the potential costs for employers and differential treatment on the grounds of seniority. In 1999, the Equality Authority was established to rule on cases of alleged discrimination. Despite the introduction of age discrimination legislation in 1998, mandatory retirement was still permitted in Ireland (OECD, 2006).

In February 2000, a preventative strategy was implemented to combat long-term unemployment. The *Action Programme for the Long-term Unemployed* and *Skills Training for Unemployed and Redundant Workers* integrated the long-term unemployed into mainstream training. Monies from the European Social Fund were used to partially finance these schemes (Department of Enterprise, Trade and Employment, 2000). The National Training Fund introduced a subsidy for employers in 2000 to allow them to run training measures within the workplace that would be accredited by the National Framework of Qualifications (OECD, 2005g). The unemployment benefits and early exit routes for older individuals were unchanged by this point.

In Ireland, older workers outside of the labour market were predominately on incapacity benefits and as a result, in 2002 a scheme was introduced called the Employment Retention Grant to aid employers in maintaining the employment of these workers (OECD, 2005g). In order to incentivise employment for those over the age of 50, new measures were introduced in 2002. Income tax exemptions were increased to over 20% for those over 65 years of age. The Back-To-Work Allowance was targeted more at certain groups of the unemployed, including the over-50s (Department of Enterprise, Trade and Employment, 2002).

In July 2004 the Equality Act was initiated in line with the EC directive which abolished mandatory retirement (OECD, 2005g). Its abolition may in some way account for the rise in the employment rate of 55–64 to beyond 50% as of 2005. Linked to the 2002 Employment Retention Grant, in 2005 a subsidy was provided for employers to promote the recruitment of older individuals with disabilities (OECD, 2005g).

From 1998, Ireland had in place the NEAP which required unemployed individuals under 25 to engage in intensive employment coaching after six months of benefit receipt; individuals over 25 were referred after 12 months which was reduced to nine months in 2000. This programme was extended in March 2003 to include a Prevention Strategy for individuals aged 25–54 after six months of unemployment which was again extended to unemployed individuals aged 55–64. From October 2006, the referral period was reduced to three months. At this point, the individual will have an appointment with a *Foras Áiseanna Saothair* (FÁS) employment officer who will determine appropriate activation measures and an individual action plan, which may include training, subsidies or referral to job vacancies (Grubb et al., 2009).

Part-time work/pension arrangements

There were no options for partial pension receipt in Ireland unless the individual earned less than €38 per week.

4.5   Portugal

The employment rate for the 55–64 year group in Portugal has remained fairly constant since the early 1990s, with the lowest level in 1993 at 45.7 and highest at 51.6 in 2003. Like Denmark, it too experienced a fall from the 2001 rate of 50.2% to the total for 2010 to 49.2%, which thus moved it just below the Stockholm target (Eurostat, 2011).

Pension policies

Pension principles and state pension age

The Portuguese pension system contained two elements: a contributory scheme that provided benefits in accordance with insurance years and a non-contributory, means-tested system. The law provides for supplementary conventional schemes as an option. In 1973, the state pension age for women was lowered to 62 years of age; the reform of 1993 (Law 329/1993) raised this threshold by six months per year until 1999 when the male and female pensions would be harmonised (Chuliá and Asensio, 2007). Thus in 1995, the state pension ages were 63 for women and 65 for men. The state pension age was not compulsory and thus individuals could continue to work (MISSOC, 1996). This reform also ruled that retirement prior to the age of 60 was not possible unless under the early retirement schemes outlined below and increased the minimum qualifying period to 15 years from ten, whilst the threshold for a full pension was raised by five years to reach 40. At the same time, the accrual rate was reduced from 2.2% to 2%, thus encouraging individuals to remain in work longer.

Law 9/99 introduced a number of changes from 2000 regarding retirement including the creation of a degree of flexibility around the state pension age (by this year, male and female pension ages were fully harmonised at 65 (MISSOC, 1999)). First, individuals could exit between the ages of 55 and 70 with a minimum contribution requirement of 30 years for early retirement. However, second, in addition to the contribution requirement, early exit would also incur a penalty of 4.5% per year short of the state pension age although individuals with more than 30 years of contributions could reduce this deduction by one year for every three years of employment beyond this threshold. Thus these changes effectively meant that a new early exit route was created for those individuals aged 55 and over with 30 years of contributions who would be subject the aforesaid reduction (Chuliá and Asensio, 2007).

Early retirement

As of 1993, the 'Anticipated Retirement' (*Reforma Antecipada*) scheme allowed those over the age of 55 in arduous work to retire early (Committee of the Regions, 2003). In addition, on health and safety grounds, certain occupations were permitted to exit early: miners (at 50 years of age); seafarers (55), flight personnel (60), air traffic controllers (55) and dancers (55) (MISSPEU, 2001). Individuals who were unemployed at the age of 60 were entitled to retire early, with a reduction of 4.5% per year.

From 2007, this had increased to 6% per year prior to the state pension age, or 0.5% per month and the age threshold was raised to 62.

Deferred retirement

Law 9/99 also introduced an incentive for deferring state pension receipt. Individuals deferring their pension receipt beyond the age of 65 would receive a bonus of 10% per year (Chuliá and Asensio, 2007; Portugal, 2000). The upper limit was 70 years of age. The Social Security Law in January 2007 sought to make the Portuguese pension system more sustainable by encouraging longer working lives, whilst at the same time ensuring pension adequacy. Thus the incentives for deferral were reformed to add 1% per month for individuals with a full contribution record (40 years); for those under the contribution threshold of a full pension, the additional amount varied from 0.33% per month for those with less than 25 insurance years and 0.65% for those with between 35 and 39 years. In addition, individuals qualifying for early retirement would receive an additional increment of 0.65% per month for deferral (OECD, 2007; Portugal, 2008).

Labour market policies

Early exit

The participation of the 55-plus has been declining since the introduction of a policy following the revolution of 1974 whereby anyone who had ever paid national insurance contributions could apply for an early pension if they had no other means of subsistence (Barros and Coelho, 1998). Thus in terms of *de facto* early exit policies, individuals could remain on unemployment benefits until the age of 60 if they were 55 when unemployed. At the age of 60, the unemployed individual could then retire early (see above). In addition, there were private early exit schemes administered by employers which were contained within the legal framework established by the 1990 Economic and Social Agreement. Under these agreements, the employer would provide a payment to the individual until they reached the state pension age (European Foundation for the Improvement of Living and Working Conditions, 2003b).

A few months after Decree-Law 9/99 was introduced (see *Pension Principles and State Pension Age*) – which created a flexible pension age and effectively produced the option to retire or exit early for those with extensive contributions – Decree-Law 199/99 was implemented which applied a contribution requirement to the *de facto* unemployment route that had allowed exit from the age of 55. From this point on, the unemployment route was restricted to individuals aged 55 and over who had been on the unemployment register for five years with a contribution record of at least 20 years. Periods spent on unemployment benefits were however counted as contribution years (Chuliá and Asensio, 2007; Portugal, 2000). Effectively, individuals who had spent 15 years in employment and five years on the employment register would have been deemed to have made sufficient contributions to access an early pension, effectively allowing them to leave the labour market at 50 years of age. However, even individuals without extensive contribution records could receive additional periods of unemployment benefits: for those older than 45, benefits could be received for 30 months plus an additional two months for every five years of contributions within the previous 20 years; for those over 55, the unemployment benefit is paid until age 60, when it is replaced by the old age pension (Committee of the Regions, 2003; MISSOC, 2000; Social Security Administration, 2006).

By 2007, the age threshold for the extended benefit duration had also been raised, in line with the limit for the early unemployment pension, to 57 and the contribution requirement had also increased to 22 years (MISSOC, 2007; OECD, 2011). In addition, the durations of unemployment benefits were not only contingent on age-contribution records also dictated the length of benefit receipt. In terms of unemployment insurance, for individuals over the age of 45 with a contribution record of less than 72 months, 720 days of payment were available; for those with over 72 months of contribution, the payment duration was 900 days, plus an additional 60 days for every five contribution years within the previous 20 (MISSOC, 2007).

Active labour market policies

In 1997, the social security contributions paid by older workers and their employers were reduced. The focus of unemployment programmes was on the young, thus leaving a gap which non-governmental organisations have sought to fill (Barros and Coelho, 1998). A key component of the Portuguese active ageing strategy was lifelong learning (Portugal, 2000) and 2000 was an integral year in terms of this policy area with many new measures initiated. Under the Adult Education and Training Courses (AET) scheme, new courses were introduced including 'Employability and Citizenship', 'Language and Information Communication Technologies', and 'Communication and Mathematics for Life'. In addition, the capabilities possessed by older workers would be valued through the establishment of Skills Recognition and Ratification Centres to award life skills with the same amount of prestige as equivalent school qualifications (Portugal, 2000).

As of 2004, a comprehensive active ageing approach was introduced in Portugal. A project entitled 'Active and Healthy Life in the Workplace' was launched as well as training for businesses to improve their health, safety and hygiene. Linked to improving individual's working lives, the option of more flexible employment was emphasised, with opportunities for working from home or part-time employment. In order to retain older workers in the labour market, Portugal promoted training and education for this group, particularly in new areas such as ICT (information and communication technology). Finally, to tackle the prejudicial views of employers and to promote the image of older workers as an invaluable resource and provide incentives for their retention, information campaigns were launched (Portugal, 2004).

The other key aim of the 2004 National Action Plan (NAP) was to promote the reintegration of those older workers already outside of the labour market. First, those older workers unemployed due to economic restructuring would take part in a programme designed to retrain individuals to re-enter new forms of work. Those who took part in these measures would receive a professional qualification. Second, the value of older workers in terms of experience was emphasised through information campaigns, apprentice schemes where older individuals would tutor the young and the promotion of voluntary work as a means to pass on these skills. In addition, the age thresholds for unemployment support and retraining were modified to allow the inclusion of those over the age of 45 (Portugal, 2004).

Part-time work/pension arrangements

There was no option of a partial pension in Portugal.

4.6   Summary and impact of reforms at the individual level

In terms of the common policy pictures shared by the nations which were close to or had already surpassed the Stockholm target when it was launched, they can broadly be divided into two sets. First, Sweden Denmark and Portugal all had early exit and retirement routes available prior to the 2000s, which were then retrenched by 2010; on the other hand, the UK and Ireland had no or very limited decommodification opportunities for older individuals throughout the entire period. Second, though opportunities to exit early were more plentiful in Sweden, Denmark and Portugal, they also provided opportunities for flexibility – both around the timing and transition into retirement – including flexible pension ages, options for partial pension receipt and incentives for deferring state pension receipt. Again, the UK and Ireland did not provide such a range of policies. Thus on the one hand, the UK and Ireland's high participation rates for the 55–64 age group could be attributed to the lack of decommodification options, whilst on the other, though Sweden, Denmark and Portugal had opportunities for early exit, they also promoted longer working lives through flexible exit options.

At the micro-level, to return to the options for de- and recommodification available to the 'model biographies', we can again see some commonality, as well as diversity. The names in the following section refer to the model biographies outlined in Section 3.6:

- the 50-plus:
  - *Laurent:* Aged 55 with 35 years of employment contributions.
  - *Jean:* Aged 55 with a disjointed employment history.
- the 60-plus:
  - *Sasha:* Aged 63 years with 35 years of employment contributions.
  - *Jude:* Aged 63 with a disjointed employment history.

In terms of this classification and Table 1.1 as outlined in the Introduction to this book regarding the entitlement conditions for policy options, Ireland, Sweden, Denmark and Portugal all had a 'rights-focused' element to the eligibility criteria for either the early exit or retirement options or both in that they were universally available to all older individuals in the 1990s. However, as they moved into the 2000s, these options began to be retrenched, either through the closure of policy options or through the imposition of contribution requirements. In Denmark where there was also a focus on contributions and therefore deservingness for some of the early retirement options, this emphasis was strengthened with the closure of the rights options; similarly Portugal moved towards a combination of rights and labour market focused decommodification to the latter and deservingness as eligibility criteria.

In the case of Ireland, a rights-focus was adopted in terms early exit eligibility and the low age thresholds for decommodification policies meant all model biographies could have been able to leave the labour market prior to the state pension age. Those over the age of 55 were able to access the early exit scheme providing they satisfied a means-test and had received 15 months of unemployment benefit. They also would have been exempt from the job search requirement of unemployment benefit receipt. These policies remained unchanged until 2007, which was comparatively late in terms of the retrenchment of decommodification options, yet a range of supply- and demand-focused ALMPs were introduced and indeed in only having one option for early exit, this was already comparatively limited when comparing EU15 nations. Thus the introduction of recommodifying policies in this case was accompanied by a retrenchment of the option for decommodification.

Sweden too initially adopted a rights-focused approach to decommodification in the third age as in the mid-1990s, the 60-plus model biographies *Sasha* and *Jude* (both age 63 with 35 years of contributions and a disjointed work history respectively) were able to decide when to exit from the labour market via the flexible retirement scheme though these choices may have been constrained by financial considerations due to the reductions (12%) to their final pension amount that would be applied. Indeed, *Jude's* disjointed work history may have prevented him or her from accruing supplementary pension credits, presenting perhaps an additional pecuniary concern. However, Sweden did have in place a citizenship-based flat-rate pension, which was comparatively generous. In addition, there was also the disability insurance scheme which allowed those over 60 to retire early (insofar as the assessment conditions were more lenient for those over 60). Individuals over this age could also partially retire if they had made ten years of contributions since the age of 45. Though the model of early retirement adopted was rights-based in the early 1990s, these rights did not extend to *Laurent* and *Jean* due to their age (55, with 35 years and a disjointed work history respectively). However, they were able to be decommodified through the unemployment benefit system which provided extended benefit durations for older individuals in 1995, though the age threshold was raised to 57 in 1997 and abolished by 2005. Over the subsequent four years new ALMPs were introduced including job creation schemes involving placements. In addition, the disability insurance scheme had been retrenched further in 1997 and no longer featured special conditions for older individuals. The early 2000s saw the modification of the flexible early pension in 2003 to increase the age threshold to 61, yet the ethos remained the same: individuals had autonomy over their exit in terms of the policy options available. In addition, the partial pension option was available for all over the age of 60. The pension system had also been modified, but there was still a basic state pension available to all citizens, regardless of contributions.

Denmark's early exit and early retirement policies presented mix of desert- and rights-focused decommodification in that there were some options available universally and some with contribution requirements. In addition, the comparatively low age thresholds meant that in 1995, the options for exit available to *Laurent* (55 with 35 years of employment contributions) and *Jean* (55 with a disjointed work history) were fairly extensive. Both would have been able to retire early if their labour market prospects were deemed to be weak (either for social reasons or related to a disability with the early social pension/anticipatory pension (*førtidspension*)) or receive unemployment benefits for an extended duration with no accompanying job search requirement. There was also an additional route available to *Laurent* due to their contribution record (the transitional income-related pension (*overgangsydelse*)). This mix is also reflected in the data for *Sasha* and *Jude* whereby the former had more decommodification options due to their contributions, including the VERP and partial pension schemes, though the latter was not completely unable to exit due to the early social pension/anticipatory pension (*førtidspension*). *Sasha* as they were eligible for the VERP scheme, could have also accessed a subsidised flexjob.

There was a significant shift in Denmark's policy approach over the following five years, as exemplified by their increased focus on participation in the labour market. Local Authorities were duty-bound to provide activation programmes before offering opportunities for exit and the individual was obliged to engage in them. In addition, the benefit extension age threshold was raised to 55 and the early social pension/anticipatory pension (*førtidspension*) was retrenched with the stipulation that those who would have been eligible should instead be enrolled on a job placement. By 2001, the age threshold for the job search exemption was raised to 58 and the benefit duration had been reduced thus limiting *Jean's* opportunity to exit the labour market. This too would have reduced *Laurent's* opportunities for decommodification through the closure of the transitional allowance (*overgangsydelse*) to new entrants after 1996. The options available to *Sasha* and *Jude* were also being curtailed with the retrenchment of the early social pension/anticipatory pension (*førtidspension*), the barring of the transitional allowance (*overgangsydelse*) to new entrants after 1996, thus making contributions key to decommodification as the only remaining option was the VERP scheme. Indeed, this scheme itself had been reformed to require 25 years of contributions in 1999 as opposed to the previous 20. As a result, for *Jude* decommodification was no longer possible unless via unemployment benefits. Incentives for labour market participation had also been introduced, including a bonus of 8,600 DKR per quarter between the ages of 62 and 65, as well as ALMPs such as the flexjob scheme, age discrimination legislation and training. Also, the state pension age had been lowered in line with many EU15 nations. The ALMPs also focused on supply-side measures such as training and education. The decommodification opportunities remained unchanged for *Sasha* and *Jude*, thus embodying an entirely desert-based system by this point. The early pension option of the VERP available to *Sasha* in 2005 had been retrenched through the increase of the age threshold to 62. In addition, as of 2010, incentives for deferring state pension receipt had increased with a tax break for those who remained in work between the ages of 60–64. The partial pensions, though reformed, continued to require a certain number of insurance years.

Portugal shifted from a mix of rights and labour market to the former combined with desert as the principle behind early exit and retirement policies. *Laurent* and *Jean*, as well as *Sasha* and *Jude* (if they were male; Portugal's unharmonised state pension ages meant women could retire at 63 in 1995) would have been able to exit early if they had been employed in certain industries or were unemployed in the mid-1990s. However, in terms of the latter, *Laurent* and *Jean* would be able to remain on unemployment benefits until the age of 60 whereas *Sasha* and *Jude* would be subject to the rules of the early retirement scheme for the unemployed which would mean their pension amount would be reduced by 4.5% per year prior to 65. The subsequent shift towards a mix of desert- and labour market-focused decommodification was exemplified through the creation of a flexible new route for long insurance durations with reductions in 1999. The routes for certain occupations were unretrenched whilst early exit through unemployment benefits was reformed to include a contribution requirement of 20 contribution years (in 1999) and limited to those over the age of 57 (2007). Thus *Jean* and *Jude* would have only been able to retire early if employed in certain occupations or arduous work; *Laurent* would in addition to these options have been able to access extended unemployment benefits until 2007 when the age threshold was raised and the flexible option with a reduction to their final pension amount. By 2010, *Sasha* would have been able to retire early with the flexible option and the unemployment pension scheme, albeit with reductions to their pension amount. In addition, ALMPs had been introduced by 2010 with a particular focus on training and skills.

The UK represents an outlier in the classification in accordance with their decommodification approach as throughout the entire ten year period, there were no early exit or retirement schemes available to any of the model biographies. The only concession to age in this respect was that the threshold for assets that determined an individual's access to means-tested unemployment benefit was higher for older individuals. However, due to the unharmonised state pension ages, *Sasha* and *Jude* could have exited at 60 years of age if female and had they remained in employment, they would have increased their pension income. If *Sasha* and *Jude* were male, access to unemployment benefits until the state pension age was again contingent on their assets, as opposed to any other criterion. The UK did however introduce ALMPs from 1996 onwards, though they tended to be limited in nature, for example age awareness campaigns as opposed to age discrimination legislation, and subsidies only once the individual entered work (indeed, only *low* paid work).

These five nations represented the vanguard in terms of participation rates for older workers in 2001 when the Stockholm target was set. In terms of the common ground between their policy approaches for older people, two sub-groups can be identified, with the UK as an outlier due to its total lack of early exit and retirement options. The first group includes Sweden and Ireland who had rights-focused decommodification options in 1995. The openness of these policies might imply that these nations should have had low employment rates for older people (acknowledging of course that there are other factors aside from social policies which affect employment rates); however this was not the case, in particular with Sweden which had surpassed the Stockholm target when it joined the EU15 in 1995 (62%, Eurostat, 2011). Nonetheless, both of these nations retrenched their decommodification options over the period from 1995–2010. Whereas other nations shifted the focus of the eligibility criteria by adding contribution requirements, Ireland and Sweden instead closed off their rights-focused options entirely at the same time as introducing a raft of ALMPs for older workers. Portugal and Denmark on the other hand did what many EU15 nations did and shifted more toward desert and contribution records as a requisite for decommodification, thus restricting the numbers of older people able to access these options. Portugal moved from a system where decommodification was available to those in certain industries and more generally to those of a certain age to focus instead on the former plus those who had made the requisite contributions; Denmark too closed off their more universal options and increased the requirements for the desert-focused options. Thus whereas in Sweden and Ireland, all older individuals would have felt the increased emphasis on 'ageing actively' through paid employment, in Denmark and Portugal those who had certain labour market histories could still become decommodified prior to the state pension age.

5

Group II: Surpassing Stockholm

The Netherlands, Finland and Germany made significant progress with regard to the Stockholm target. In 2001, all were below the 50% employment rate goal but within nine years exceeded it at a rate that was above the average for all EU15 nations (9.6%) (Eurostat, 2011). Again, as with the preceding chapter, the policy context prior to 2001 will be addressed, as well as the reforms undertaken until 2010. The impact of these changes at the micro-level will then be explored and compared.

5.1   The Netherlands

The Netherlands was initially in the EU's Group III category for nations who were close to the EU average for 55–64 employment rates yet below the Stockholm target (Commission of the European Communities, 2004: 8). However since 2001, these rates have improved dramatically. In the early 1990s, the employment rate for this age group was around 28% but began to steadily increase from the latter part of this decade onwards to exceed the Stockholm target in 2007 (Eurostat, 2011).

Pension policies

Pension principles and state pension age

The Dutch pension system consisted of three elements. The basic old age public scheme (AOW) bestowed equal pension rights on all citizens and could only be accessed from the age of 65. This basic pension was based on years of residence (50 years between the ages of 15 and 64) as opposed to contributions (OECD, 2005j; Bertelsmann-Stiftung Foundation, 2005a). The final pension amount was equal to the net minimum wage. This pension was not means-tested and a supplement existed for those with no other form of income (OECD, 2005j).[1] The second pillar, the collective occupational scheme, was funded by employers and employees and was akin to an individual fully-funded pension. These pensions were controlled by a legal framework through the Pension and Insurance Supervisory Authority (PVK) but were managed by the social partners. Of all EU15 nations, the Netherlands has the most developed occupational pillar (OECD, 2005j) as it was compulsory for most employees and the amount corresponded to years of insurance and level of income (de Vroom, 2004b). As a result of the gendered division of labour, 37% of women were excluded from these schemes compared to 10% of men. In 1985, these pensions made up over one-quarter of individuals' retirement incomes (Walker, 1993), thus the absence of access to such a scheme had a significant impact on post-retirement income. Finally, the private pension element was distinct from the state and the collective schemes (OECD, 2005j; Bertelsmann-Stiftung Foundation, 2005a).

Early retirement

There were three main early exit routes from the labour market: the early retirement pension (*Vervroegde Uittreding*, VUT), the occupational disability programme (*Wet op de Arbeidsongeschiktheidsverzekering*, WAO) and unemployment benefits (*Werkloosheidswet*, WW); only the latter two were directly provided by the state. The early retirement pension (VUT) and the disability pension (WAO)[2] played the largest roles in the declining rates of older individuals in employment; the unemployment benefit option was the least popular among this group. In terms of the WAO, early retirement was granted to those whose reduced working capacity meant they could not earn the same as healthy workers with the same or similar training. It could be claimed at any age, though those aged 53–57 could access the benefit for three years; for those over 58 for six years and those over 59 could remain on the benefit until the age of 65 when they could claim the state pension. However, the degree to which the WAO scheme can be classed as an early exit option is contentious as though it was open to all ages, special additional periods applied to older individuals. Similarly, special conditions also applied to unemployment benefit recipients over the age of 57.5, thereby providing a *de facto* early exit route (see *Early exit*).

The VUT schemes were separate from the pension system and were financed by employers and employees on a pay-as-you-go principle and thus there was variation between companies. However, the state still exerted significant influence over the scheme through fiscal policies and "confirmation… [whereby] the state has the ability to 'legalize' exit arrangements, as part of a collective agreement between the unions and employers' associations by declaration" (de Vroom and Blomsma, 1991: 112). De Vroom and Blomsma highlight the distinct nature of the VUT systems however, as separate from state-provided schemes which are defined as a "public (collective) commodity for the working population provided by the government" (ibid: 112).[3]

Due to the aforementioned early exit trend and low female employment, projections of a labour market supply shortage created concern. As a result, "[s]ince the second half of the 1990s the 'early exit group' was reframed from a 'solution' into a 'problem'. In policy documents the early exit group suddenly appeared as a 'marginal group' which should be a target for reintegration policies" (de Vroom, 2004b: 134). Five groups were targeted for special active employment policies, including older individuals in addition to people entitled to unemployment and social assistance benefit, people who were disabled, non-native persons and women (de Vroom, 2004b). Three approaches were introduced to reverse the early exit trend. First, the Dutch government reorganised the social security system to curtail access to disability pensions. Second, preventative measures were introduced in the form of age awareness campaigns for employers. Third, an attempt was made initially to integrate older workers into the labour market with targeted measures which were usurped in favour of subsidised employment (Klercq, 1998). The latter two reform areas will be discussed below in the section on active labour market policies. Of those exiting between the ages of 50 and 64 in 2002, 13% were on disability benefits, 6% had accessed early retirement schemes, 3% were on unemployment benefits and 3% were on social assistance (OECD, 2005j). Thus disability benefits were one of the main early exit routes in the Netherlands and as a result, in 2002 the Gatekeeper Act was passed to make the screening process for disability benefits more rigorous with periodic re-assessments to monitor the individual's condition.[4]

Perhaps to ameliorate the effects of the increased conditionality related to early exit through unemployment, the WAO was reformed in 2004 to include a reassessment requirement, from which those over the age of 55 were exempted "since these persons have insufficient chance of reintegration due to their relative distance from the labour market" (Ministerie van Sociale Zaken en Werkgelegenheid, 2004: 23). The WAO was further reformed in 2006 to be replaced by the *Work and Income (Capacity for Work) Act* (Werk en Inkomen naar Arbeidsvermogen, WIA). Those already on WAO were subject to re-assessment whilst the scheme was closed to those who became disabled from January 2004 who were instead subject to the WIA Act. At its core is the emphasis on encouraging employers and individuals who are partially disabled to consider employment. If individuals who are partially disabled remained on benefits, they received a benefit based on the minimum wage; for permanently and fully disabled individuals, the benefit amount was 70% of their previous wage under the Income Support Scheme for Persons Incapable of Work (IVA). If the partially disabled individual entered employment, they would receive a subsidy amounting to 70% of the difference between their new and former wage as part of the Work Resumption Benefit for Persons Partially Capable of Work (WGA) (Bertelsmann-Stiftung Foundation, 2010).

Deferred pension

There are no incentives for pension deferral in the Netherlands, but it is possible to combine work and pension receipt post-65.

Labour market policies

Early exit

In the 1990s, the unemployment benefit duration in the Netherlands corresponded to the individual's contribution record yet individuals over the age of 57.5 and unemployed were exempt from the job search requirement of unemployment benefit receipt. By 2000, the unemployment benefit schemes had been reformed to include three elements: the short-term benefit (*kortdurende uitkering*), the salary-related benefit (*loongerelateerde uitkering*) and the follow-up benefit (*vervolguitkering*). All three had a contribution requirement and the latter, the follow-up benefit allowed individuals over the age of 57.5 who fulfilled 26 weeks of employment within the previous 39 and four of the five previous years (working at least 52 days per year) to remain on unemployment benefits until they reached the state pension age. However, the general job search exemption remained unreformed.

Measures were undertaken in the Netherlands to prevent long-term unemployment becoming an automatic pathway to early exit. Social Assistance has been a right in the Netherlands since the 1960s in that the Dutch government until 2004 was duty-bound to assist individuals in their job search or, if work was not available, to provide benefits. These responsibilities were laid down in the *Law on General Social Assistance* (*Algemene Bijstandswet*, ABW), the *Law on the Retaining of Job-seekers* (*Wet inschakeling werkzoekenden*, WIW) and the *Decree on In-Flow Jobs* (*Besluit in- en doorstroombanen, ID-besluit*). From 1982, older individuals had been exempt from the job search requirement of unemployment insurance or social assistance if they were 57.5 years old and over (OECD, 2005j). In 2004, a new law, the *Work and Social Security Act (Wet Werk en Bijstand)* superseded the aforementioned laws,[5] removing additional unemployment benefit for those unemployed after 10th August 2003 and refocusing the emphasis on the responsibility of the individual to engage in the labour market. This provided a benefit set at a maximum of 70% of the minimum wage to individuals aged 57.5 and over who could then receive this benefit for 3½ years after the general unemployment benefit had expired (at between four to five years) (OECD, 2005j). Also as those over 57.5 were no longer exempted from actively seeking work and they thus had to accept any 'generally accepted employment' offered to them. However, the Work and Social Assistance Act gave municipalities discretion to exempt certain individuals aged 57.5 to 64 years of age who were deemed to have little chance of re-employment. In addition, older individuals on WAO were still exempt from job search (OECD, 2006; Committee of the Regions, 2003).

Active labour market policies

The Netherlands, in comparison with other EU15 nations, acted early in terms of anti-age discrimination measures. Article 1 of the Dutch Constitution (1983) states that "everyone in the Netherlands should be treated equal. It is prohibited to discriminate against persons on the basis of their religion, political preference, race, sex, or any ground whatsoever" (Bertelsmann-Stiftung Foundation, 2000b). In 1993 the Association Against Age Discrimination was established to promote legislation against age discrimination and in 1994 National Office against Age Discrimination was set up (Taylor, 1998). The Integral Programme of Action for Policy and the Elderly created three guidelines on age discrimination to run from 1995 to 1998. First, age discrimination was clearly defined as "a situation in which the use of age to discriminate cannot be justified" (de Vroom, 2004b: 136). Second, any age limits had to be legitimised and finally, the precedent was set that this legislation was "a tool to defend the elderly against physical and psychological declination is still a necessary regulative criterion in law making and is seen as an indication of the 'level of civilisation of a society'" (de Vroom, 2004b: 136). New guidelines for redundancies were introduced in 1995 to alter the selective dismissal rules to 'last in, first out' in the case of economic difficulties and to ensure older workers did not make up a disproportionate number of redundancies by ensuring dismissals mirror age compositions (OECD, 2005j). However, in spite of this apparent anti-ageist stance, in the early 1990s, state subsidies for employment did not apply to those over the age of 57.5 (Drury, 1993). In addition, there was no minimum wage requirement for those over the age of 65 and dismissal arrangements were different from the rest of the working population with Dutch courts ruling that these individuals had less of a need of work due to their pension entitlement (Drury, 1993: 49).

As part of the aforementioned tripartite strategy to combat early exit, the Netherlands continued its work on anti-ageism policies with the launch of an action campaign in 1996 by the National Office Against Age Discrimination. Though court rulings prohibited contracts that included clauses for early exit, the anti-discrimination Article was open to some degree to interpretation, leading to a 'hit and miss' series of challenges to ageism (Taylor, 1998). Furthermore, a Bill was introduced in 1997 on age discrimination which was later replaced in 1999 with a more encompassing piece of legislation to cover training, education, promotion and recruitment. However, age limits could be retained in order to promote health and safety, to maintain a balanced age composition of the workforce and if age was an important factor in a particular job (de Vroom, 2004b).

The strategy to promote later exit in the late 1990s also included targeted activation measures and subsidised employment. The Netherlands removed the age thresholds for state subsidies for employment in 1998 (which had been 57.5 years of age) (Drury, 1993). Measures exclusively targeted at older individuals from minority ethnic groups were introduced, acknowledging the different barriers they face (Taylor, 1998). In addition, self-help groups were launched. This final policy was abandoned in favour of subsidised employment or '*Melkert banen*' (Klercq, 1998). In terms of training, pre-retirement courses were introduced in addition to schemes where older workers were utilised as consultants. In 1998, an experience rating was applied to disability benefits (OECD, 2005j).

In July 1999, individuals who took up employment with lower wages or working hours retained their right to unemployment and disability benefits based on their previous wage levels (Committee of the Regions, 2003). As of July 1999, legislation prohibiting age discrimination was introduced to which employers could apply for exemptions (Committee of the Regions, 2003). On 14th June 2000, in accordance with Article 13 of the EC Treaty the *Wetsvoorstel verbod op leeftijdsdiscriminatie bij de arbeid* (Proposed Law on Banning Age Discrimination at Work) was launched which banned differential treatment according to age in terms of training, recruitment and promotion (Delsen, 2002). The 2001 Taskforce on

Older People and Employment for employers was created to run an information campaign to alter employers' perceptions of older workers until 2007 (OECD, 2005j).

The new law of 2004, the Work and Social Security Act (*Wet Werk en Bijstand*) superseded these laws (Bertelsmann-Stiftung Foundation, 2004a), removed special assistance for the unemployed and instead all unemployed individuals were eligible for a tailored 're-integration strategy' and those between 45 and 54 years of age were particularly targeted. The tax deduction introduced in April 2004 was also designed to make work more attractive to older workers whilst VUT was no longer tax deductible to encourage longer working lives (OECD, 2005j). From 2004, to encourage unemployed individuals to take up employment at a lower wage than they had previously received, in the event of multiple spells of unemployment, their benefits would be based on the highest of these salaries (OECD, 1997, 2005j).

To encourage the recruitment of those unemployed over 50 years of age and the retention of those over 55, employers were completely or partially exempted respectively from paying part of their disability insurance contribution as of 2004 (OECD, 2005j). This year also saw the Life Course Bill come into effect. Under this Bill, the Life Course Savings Scheme was introduced to promote gradual exit from the labour market. As a result, an older individual could reduce their working time by 50% in the two years before retirement. The individual could not utilise this scheme to exit early and thus would remain engaged in the labour market to some extent until they reached state pension age (Ministerie van Sociale Zaken en Werkgelegenheid, 2003). In 2004 the Ministry of Social and Employment Affairs (SZW) launched a subsidy for employers wishing to pilot ways of retaining and employing older individuals. The subsidy was available from 2004 to 2007 with €21 million allocated to it (Bertelsmann-Stiftung Foundation, 2005a). From January 2004, employers were obliged to cover the unemployment benefit of older workers they made redundant (Ministerie van Sociale Zaken en Werkgelegenheid, 2003).

In May 2004, a new law on age discrimination was passed regarding equal treatment in employment (OECD, 2005j). The Dutch Equal Treatment Act on the basis of Age in Employment (WGBL) forbade discrimination on the grounds of age in recruitment, training and dismissal unless it could be objectively justified. Somewhat paradoxically, two years later in March 2006, the 1995 guidelines on the dismissals of older employees which had opted for a 'last in, first out' approach were revised to spread redundancies across whole workforces equally, thus reducing the protection older workers had previously enjoyed (Bredgaard and Tros, 2006).

Part-time work/pension arrangements

Partial retirement is not available in terms of state-provided pensions in the Netherlands. However, in July 2000 the *Wet aanpassing van de arbeidsduur* (Law on the Adjustment of Working Time) was introduced to allow all workers to request to reduce their working hours following one year of employment with their firm (Delsen, 2002). Depending on the collective agreement, workers aged 58–60 may be able to reduce their working hours by 1–2 days a week (Van Oorschot and Jensen, 2009).

5.2   Finland

Finland had already begun to increase the employment rate for older individuals by the time the Stockholm target had been set, having experienced a drop in the early 1990s. From 2001 to 2010, the rate had increased by 10.5% from 45.7% to 56.5% (Eurostat, 2011).

Pension policies

Pension principles and state pension age

The Finnish pension system pre-1996 provided a universal National Pension with earnings-related supplements used to 'top-up' this amount. In instances where no contributions had been made to the earnings-related element, the state would pay the supplement. The pension age was 65 (Kangas, 2007). 1996 saw a major reform of the Finnish pension system whereby the National Pension became subject to income-testing. As a result, there was increased focus on the employment pension (*Työeläke*) which covered all economically active persons and required 40 years of contributions for the full pension amount (which in the case of the employment pension amounted to 60% of the pensionable salary). The universal National Pension scheme required 40 years of residency and income below a certain threshold (Kangas, 2007). Reforms actualised in 2005 created a flexible pension age between 63 and 68 with different accrual rates (see *Early retirement* below).

Early retirement

As will be explored in the section on *Early exit*, the unemployment tunnel in Finland had been in place since the 1970s and fed into the unemployment pension (*Työttömyyseläke*) whereby at 60, individuals could receive this early pension until state pension age if they had made 15 years of insurance contributions and exhausted their right to unemployment benefits (OECD, 2005d; the age limit has been reformed many times and was decreased to 58 in 1978 and further to 55 in 1980 before being gradually raised back to 60 after 1986). Thus individuals could effectively exit permanently from the labour market at 53 through unemployment benefits and the unemployment tunnel, before entering the unemployment pension at the age of 60 (OECD, 2006). In addition, there was a pension available for farmers from 1974 who could retire between the ages of 60 and 64. Alternatively, those who converted to part-time work or transferred their farms to successors for non-agricultural purposes could exit at 58 years of age (MISSOC, 1996).

As a result of these early exit and retirement policies, the participation of older workers in Finland declined in the 1990s, resulting in an average retirement age of 58 despite the fact that the state pension age stood at 65 (Tikkanen, 1998a) and by the mid-1990s, only one third of people aged 55 to 64 were employed. By the late 1990s, this situation had yet to improve. In 1998 the employment rate for workers aged 55 to 64 stood at 33.9% (Tikkanen, 1998b). A year later, the effective retirement age stood at 59 (OECD, 2005d). Gould and Saurama (2004) argue this was due to the expansion of the unemployment allowance, providing a route out of the labour market for these individuals.

In 1986 the individual early pension (*Yksilöllinen varhaiseläke*) was introduced for those in the private sector between the ages of 55 and 64 with an extensive contribution record with their current employer. Individuals were able to exit via this route in cases where illness or disability, years of employment, ageing factors or exhaustion meant they could not continue in the labour market (European Foundation for the Improvement of Living and Working Conditions, 2003a). The amount of pension received corresponded to their age, contribution record, working conditions and health. If the individual later returned to work, the pension was either halved or suspended (OECD, 2005d). This was distinct from the disability pension (*Työkyvyttömyyseläke* – not included here as it is open to those aged 16–64) as the exhaustion criteria were less specific. This scheme was further expanded to include the public sector in 1989. Those in the public sector could exit at 58 and both sectors would incur a reduction of 6% to their pension amount for every year they were under the age of 65. In addition, an early old age pension (*Varhennettua Vanhuuseläkettä*) was introduced for those over the age of 60 with the same level of reductions.

The recession meant that Finland began to turn away from early exit policies in the late 1980s but the concrete changes were not enacted until the 1990s. Changes to the early exit routes in Finland were gradual for a number of reasons. First, the public support for these programmes was extensive. Second, a number of path dependencies were in place, with policies locked into a set course that could not be dramatically changed due to the nature of pension rights and benefits. Third, the high rates of unemployment in the 1990s made the retrenchment of early exit difficult as the other recourse open to recipients would largely be unemployment benefits. Linked to this, the high proportion of youth unemployment meant "older employees felt that they had not just a moral right but a moral duty to retire early in order to make room for the young in a tight labour market" (Gould and Saurama, 2004: 75). This high unemployment persisted even once Finland had emerged from the recession. Finally, the early exit policies were formed following negotiations with social partners and thus large-scale reform was slow (Gould and Saurama, 2004).

Between 1995 and 1997, the age threshold for the individual early pension for individuals working in the private sector was raised to 58 (Hytti, 2004). Autumn 1999 saw the passage of a Bill through parliament regarding pension reform which subsequently came into force in 2000. The reform covered the unemployment pension system and individual early retirement pension scheme, but retained the 'unemployment tunnel' for those aged 55 and over, allowing them to remain on unemployment benefits beyond the maximum 500 day period until they reached the age of 60 when they could receive the unemployment pension (the age threshold had been raised from 53 in 1997 – see *Early exit*). The unemployment pension previously also included a clause called the 'Right to the Future Time' which meant that the pension amount would consist of the individual's contributions, had they been previously employed. These reforms removed the latter clause to reduce the pension amount by 4% at most in relation to the wage rate. This pension tunnel scheme was attractive to employers who only had to fund 50% of the pension if their firm employed more than 300 people. This amount was raised in this year to 80% to match the required funding of disability pensions (Bertelsmann-Stiftung Foundation, 2001b). Thus from 2000, the amount of the unemployment pension was reduced by up to 4% per year. This measure was designed to discourage early exit, yet at the same time, the conditions attached to this benefit had been reduced in 1998 to remove the contribution requirement (Finnish Ministry of Labour, 2001).

In addition to these amendments, the individual early retirement pension scheme for those with a diminished working capacity was modified in 1999. The age threshold for this scheme was raised to 60 unless the individual had been born before 1944 (Vinni, 2002). The logic behind this change was to encourage the retention of older workers and make the unemployment pension route less attractive to employers and employees alike. Conversely, from 2000 the reduction for the receipt of an individual early pension and an early old age pension was cut from 6% per annum to 4.8% or 0.4% per month (MISSOC, 2000). From 2000, employees were able to take leave from work for a specified period, providing an unemployed individual filled the vacancy. Similarly, there was a part-time subsidy scheme which allowed an individual to reduce their working time by half, so long as an unemployed individual was recruited to make up the full-time hours (Committee of the Regions, 2003). In order to encourage employers to retain older workers, in 2000 the cost of disability and unemployment pensions were shifted from the state to companies (Gould and Saurama, 2004). As of 2000, the increase for those who deferred their pension was reduced from 1% to 0.6% (MISSOC, 2000).

In March of 2001, a commission was established to discuss possible directions for the pension systems. As part of this blanket of reform, the private sector engaged in dialogue and agreed with the proposed changes. In November 2001, an agreement was reached with the social partners to increase the effective retirement age by two to three years and the proposed changes were put forward in Autumn 2002 and passed through Parliament in 2003 to become effective in 2005 (OECD, 2005d). These reforms were designed to increase the effective retirement age by two to three years (Ilmakunnas and Takala, 2005; Lassila and Valkonen, 2006). Thus the Finnish pension system underwent reform to improve the organisation of its funding (Bertelsmann-Stiftung Foundation, 2003b). As part of these reforms, individuals would accrue pension rights from the age of 18 (as opposed to the previous arrangement of 23) and a flexible retirement system would be introduced. Individuals would be able to retire between the ages of 63 and 68 as opposed to the fixed threshold of 65. This flexibility was managed by providing differential accrual rates according to age so as to incentivise longer working lives. The rate for those aged 18 to 52 was set at 1.5% per annum; for those aged 52 to 62, it was set at 1.9% per annum (an increase on the previous 1.5%); for those aged 63 to 67, it was 4.5% per annum (this was previously 2.5%); and for those over 68, the accrual rate was set at 0.4% per month or 4.8% per year. In addition, those exiting at 62 received a reduction of 0.6% per month under the age of 63. Thus individuals exiting at 62 had their final pension amount reduced by 7.2% in addition to not receiving the 4.5% bonus for working the additional year. This sharp increase post-62 was argued to provide an incentive to prolong labour market participation (OECD, 2005d). The final pension amount would be calculated in accordance with lifetime earnings as opposed to the previous system whereby the average of the last ten years was utilised. The ceiling of 60% was also abolished under this reform (Finnish Ministry of Labour, 1998; OECD, 2005d, 2006).

With regard to the early retirement schemes, in 2005 the individual early retirement pension was closed to new entrants and the age threshold for the early old age pension was raised to 62. To ameliorate the effects of this change, the assessment for disability pension was relaxed for those over the age of 60 (Finnish Ministry of Labour, 1998, 2005; OECD, 2005d). The farmers' pension remained unreformed from 1995. In addition, the unemployment pension scheme is to be abolished gradually between 2009 and 2014 in that it was made only available to those born before 1950 (Social Security Administration, 2006; Hytti, 2004). Individuals who would have previously been eligible for this scheme would instead access unemployment benefits until the age of 65, which employers would be jointly liable for (OECD, 2005d; Hytti, 2004). With the abolition of the unemployment pension and the redirection of its clients to extra unemployment daily allowance entitlement, older individuals were placed under the remit of the unemployment insurance system as opposed to the pensions' element (Gould and Saurama, 2004). The changes to the various early retirement schemes are summarised in Table 5.1.

Deferred retirement

Those deferring their pension beyond the age of 65 received an increase of 1% per month above this age in the mid-1990s, up to the age of 68 (MISSOC, 1996). As of 2000, the increase for those who deferred their pension was reduced from 1% to 0.6% (MISSOC, 2000). The changes,

Table 5.1 The Finnish pension system pre- and post-2005

	Target groups	Pre-1995	1995	2000	Post-2005
National pension (*Kansaneläke*)/Statutory earnings-related pension (*Työeläke*)	All workers	65	65	65	62/63–68
Early old age pension (*Varhennettua Vanhuuseläkettä*)	Older workers	60–64	60–64	60–64	62+
Disability pension (*Työkyvyttömyyseläke*)	Disabled	16–64	16–64	16–64	Relaxed rules for people aged 60+
Individual early retirement (*Yksilöllinen varhaiseläke*)	Sick and aged	55–64 (private sector) 58–64 (public sector) 6% reduction	58–64 for both sectors	60–64	Closed to new entrants
Part-time pension (*Osa-aikaeläke*)	Part-time workers	60–64	58–64	56–64	58–68
Unemployment pension (*Työttömyyseläke*)	Long-term unemployed	55–64	55–64	60–64	Phased out 2009–2014
Unemployment tunnel (*Työttömyyseläke*)	Long-term unemployed	53–65	53–65	55–65	57–65
Farmer's pension	Agricultural workers	55/58/60–64	55/60–64	55–64	55–64
Deferred old age pension	Older workers	65	65	65	68

Sources: Adapted from Uusitalo, 2007; OECD, 2005d: 77.

which were actualised in 2005, meant the state pension age was more flexible and the different accrual rates were designed to provide an incentive to remain in the labour market. For those aged 63 to 67, accrual was 4.5% per annum (this was previously 2.5%) and for those over 68, the accrual rate was set at 0.4% per month or 4.8% per year.

Labour market policies

Early exit

By the mid-1990s, Finland had a great many early exit routes. Gould and Saurama (2004) argue Finland's first phase of early exit policy introduction in the 1970s to 1980s saw the alteration of existing social security arrangements, such as those for disability[6] and unemployment to allow individuals to exit early from the labour market. With regard to the amendment of unemployment policy, Gould and Saurama argue the seeds for later early exit policies were sown in 1971 with the creation of an 'unemployment pension device' (*Työttömyyseläkeputki*) or 'unemployment tunnel' which comprised of an extended unemployment benefit duration and unemployment pension for the long-term unemployed. At this point, unemployment was only 3% and thus the policy was created to provide exit for those (relatively) few individuals with limited re-employment prospects. The unemployment benefits system in Finland included an unemployment allowance which is comprised of the basic allowance paid by the state through the Social Insurance Institution (UA) and the earnings-related allowance paid by the Unemployment Insurance Funds (UI). Generally, those receiving unemployment benefits had to fulfil a 'period of employment condition' (*Työssäoloehto*) of having worked for 26 weeks during the previous 24 months. In turn, the maximum benefit duration for either the earnings-related or basic benefits was 500 days, after which six months of employment was required before benefits could be accessed again. Those who do not meet the employment criteria or who already received unemployment allowance for 500 days could receive a means-tested labour market subsidy from the state for an unlimited period (Koskela and Uusitalo, 2003). The unemployment tunnel allowed individuals who turned 53 years and one month before the benefit duration elapsed an extension until the age of 60 when they could receive the unemployment pension (*yöttömyyseläke* – see *Early retirement*). Thus the combination of existing unemployment benefits and the tunnel scheme effectively allowed individuals to exit at age 53 years and one month. Though there was no *de jure* job search exemption, the unemployment tunnel acted as a *de facto* ruling (OECD, 2005e; Uusitalo and Koskela, 2003).

In 1997, the 'unemployment tunnel' was modified to increase the lower age limit of the extension of benefit to 55 and one month (Uusitalo and Koskela, 2003; Roschier Attorneys Ltd., 2007; Vinni, 2002). The eligibility criteria for unemployment benefit receipt had also changed to include an initial condition of a minimum of 43 weeks (with a further minimum of 18 hours a week) of employment during the previous 28 months. In 2005, the age threshold for the unemployment tunnel was raised by a further two years to 57 (Finnish Ministry of Labour, 1998; OECD, 2005d).

Active labour market policies

Comparatively, Finland began to address the challenge of active ageing at an earlier stage than other EU15 nations. In attempting to reverse the early exit trend, the Finnish government acknowledged that the tightening of eligibility criteria for early exit schemes would not reduce the number of applications, albeit making them less successful; the prospect of continuing in the labour market needed to be made more attractive. Particular attention was focused on ill-health and working conditions as the mental and physical strain of previous employment was the second most cited reason for early retirement (Tikkanen, 1998a). From 1995 to 1999, those aged 50 to 58 were the special focus of active labour market measures whereby they would be called to an interview at the job centre from which a tailor-made action plan was created. Individuals who refused to participate would face sanctions (Vinni, 2002).

Following a resolution passed by the Council of State on 6th February 1997, the 'National Programme on Ageing Workers' was established to run from 1998 to 2002 with a budget of €4 million (Vinni, 2002; Mandin, 2004). The spheres of intervention included the individual level, the workplace, the labour market and 'system factors' including pensions and labour market legislation. The programme included 40 measures encompassing training, information campaigns emphasising the benefits of employing older individuals and measures to improve working capacity along the programmes three main themes: developing working life, promoting the return to work and modifications to the pension and social security to promote employment. Certain measures were carried forward from the programme to run until 2007–9 including the *Veto* scheme to improve health and safety at work; *Noste* which focused on training; *Kesto* to improve functional capacity and make work more attractive; *Tykes* to promote cases of best practice and *Kuntatyö* to improve working conditions in the public sector (Taylor,

2005; Piekkola, 2008). Temporary legislation was introduced to allow an unemployed person over the age of 55 to accept a non-permanent, low-wage job and not face a reduction in their pension amount (Finnish Ministry of Labour, 2000; OECD, 2005d). In 1998, a scheme was introduced to provide employment subsidies for a maximum of 12 months for the long-term unemployed (Committee of the Regions, 2003).

Age discrimination legislation was also included in the adjusted Finnish Constitution (731/1999, as amended) which contained equality as a principle and therefore outlawed discrimination in every form, including on the grounds of age (Roschier Attorneys Ltd., 2007). 2001 saw many measures to improve the employability of older workers through training initiatives. New vocational training schemes were launched aimed at those over the age of 40. Finland's anti-ageism legislation was based on the ILO convention on equal treatment in work which was extended in 1987 to include recruitment. This was strengthened with the Employment Contracts Act (55/2001) in 2001 to oblige employers to treat employees and job seekers equally with regard to recruitment and the advertisement of vacancies (OECD, 2005d). Discrimination on the grounds of age is prohibited by the constitution and by the Employment Contracts' Act (Committee of the Regions, 2003).

The most common motive for early retirement in Finland is disability. As a result, in 2004 before an individual would be granted access to the disability pension, vocational rehabilitation had to be considered. Measures included in this rehabilitation included career guidance, work experience placements, training or financial help in order to become self-employed (Finnish Ministry of Labour, 2005).

Part-time work/pension arrangements

The part-time pension (*Osa-aikaeläke*) was also introduced in the late 1980s and different rules applied to those in the public sector (available from 1989 for those aged 58 and over) and private sector (available from 1987 for those aged 60 and over) (Committee of the Regions, 2003; Belloni et al., 2006). Under this scheme, individuals could work between 35–70% of their previous working hours (to 16–28 hours per week) and receive their pension set at 50% of the difference in income to supplement the new wages. As of 1994, the ages to the private and public sectors were harmonised to 58. To make the combination of part-time work and pensions more accessible, the scheme was modified in July 1998 to run until the end of 2002 whereby the age at which an individual could combine part-work and pensions was lowered to 56 years. As of 2003, the age threshold was increased again to 58 (Finnish

Ministry of Labour, 1998; Gould and Saurama, 2004; Hytti, 2004; Belloni et al., 2006).

5.3   Germany

Of all EU15 nations, Germany made the most significant gains in terms of the employment rate of older individuals, increasing by 19.8% from 37.9 in 2001 to 57.7 in 2010 (Eurostat, 2011). This rise came relatively late, with the employment of older individuals remaining under 40% until 2004, after which it increased significantly. However, though this would imply that policy began to shift towards the active ageing agenda in the early 2000s, in actuality moves began in the early 1990s with the Pensions Reform Act 1992 which introduced reforms which would only come into effect in the following decade, as will be explored below.

Pension policies

Pension principles and the state pension age

Since 1957, the German Pension system has operated on a PAYG principle. Contributions are compulsory unless the individual was a trainee, disabled or in military service. The minimum period of membership is five years (Bertelsmann-Stiftung Foundation, 2000a) and the standard age for retirement was 65 years, though under certain conditions, women could exit at 60 (see *Early retirement*; Schmähl, 1998). The pension age will rise from 2012 to 2029 from 65 to 67 (Social Security Administration, 2008).

Early retirement

There were two main early old age pensions (Teipen and Kohli, 2004), introduced in 1957 (OECD, 2005f). First, individuals could exit at the age of 60 if female with 15 insurance years of which ten where accrued over the age of 40. Secondly, if long-term unemployed, individuals could also exit at 60 if had they been previously employed for 15 years. The former was designed to compensate for the shorter employment histories of women and the use of the latter increased greatly in conjunction with the reform of the unemployment benefit system in the late 1980s (see *Early exit*; OECD, 2005f).

During the 1970s, Germany faced persistent and high unemployment and the policy response involved reducing the labour supply. The liberal-left coalition introduced a flexible seniority pension (*flexibles Altersruhegeld*) in 1973 which allowed individuals to exit at the age of 63 with 35 years of contributions (Teipen and Kohli, 2004; Ebbinghaus, 2006). There was also a disability element to this scheme which allowed individuals with 35 years of contributions and a severe disability (defined as a loss of earnings capacity by 50%) to exit at 62 years of age, which was then lowered to 60 after 1978 (Börsch-Supan and Jürges, 2011).[7] In addition to the loss of earning capacity, the definition of '*Erwerbsunfähigkeit*' (incapacity to work) also encompassed the situation in the labour market (Mandin, 2004; Guillemard, 1993).

Early exit was an area of consensus in Germany in the 1980s with its proponents arguing it increased employment opportunities for the young. To ameliorate rising costs for the state, Germany shifted the payment of early pensions to employers, thus providing less of an incentive to shed the oldest of the workforce (Taylor, 1998). The 1984 Early Retirement Act (*Voruhestandsgesetz*) regulated agreements between Trade Unions and employers for compensation for those who exited voluntarily as early as 58 years of age. This Act was in place from 1984 to 1988 and allowed workers to retire at 58, providing their employer granted payments until the individual reached the state pension age. This stipulation was designed to prevent the older individual moving to disability benefits and thus reduced the cost for the state (Teipen and Kohli, 2004).

In November 1989, Pensions Reform Act 1992 was enacted by the German Parliament. This Act represented a large-scale reform of the pension system in Germany and the changes it introduced came into effect in 2001, with a view to closing all early retirement benefits except those for disabled individuals by 2012. The principle behind this reform was that as of 2001, the economic situation would be more favourable and therefore early retirement could be retrenched (Mandin, 2004). From 2001, the earliest men and women could exit from the labour market was 62, with reductions of 3.6% per year prior to the state pension age of 65. The disability element of the long insurance duration pension was exempt from this deduction (Schmähl, 1998; Hinrichs, 2005).

Following the reforms of 1992, further changes were made to deter older individuals from exiting at 60 via the early retirement route for unemployed persons. In 1996, the federal government took steps to accelerate some of the reforms of the 1992 Act, which were due to come into effect in 2001. First, the reductions for claiming an early pension in the case of unemployment were imposed as of 1997, as opposed to by 2001 as outlined in the Pension Reform Act. The age threshold for this pension was also increased by three years by 1999 to reach 63, and it would then increase to 65 by 2001 (Mandin, 2004; Schmähl, 1999). From 1999, the age threshold for the 'flexible' early pension would also increase to reach 65 by 2001 for those born after 1952. The age threshold for the early retirement scheme for women was also raised to reach 65 by 2004, but this process began in 2000 (Schmähl, 1999; OECD, 2000; Federal Republic of Germany, 1998; Hinrichs, 2005).

In 1997, the 1999 Pension Reform Act was proposed which increased the threshold for early retirement in the case of disability from 60 to 63. In addition, the disability scheme also incurred a reduction of 3.6% per annum, with a maximum of 10% (OECD, 2006). The 1999 Pension Reform Act further tackled early retirement by gradually increasing age thresholds and initiating the closing off of pathways by 2016. As such, from 2012, the earliest an individual could retire would be 62 years of age with 35 years of contributions (OECD, 2005f; Schmähl, 1999).

In 2004, the Old-Age Pensions Insurance Sustainability Act (*Rent-enversicherungs-Nachhaltigkeitsgesetz*) altered the existing early retirement schemes arguably to take into account demographic trends so as to ensure the sustainability of pension arrangements (OECD, 2005f). As of 2004, the age threshold for all pensions was raised to 65 for certain birth cohorts. The pension schemes for women and the unemployed were restricted to cohorts born before 1952, thereby effectively creating two early retirement systems for different birth cohorts. The older birth cohorts were subject to the aforementioned rules and reductions that were gradually being introduced from 1999, whilst for the younger cohort, the early pension for women had been abolished, the age threshold for the long-term insured was being increased to reach 65, and in terms of disability, the age threshold was increased to 63 for a full pension (those exiting before this age would incur a reduction of 0.3% per month). As a result of these reforms, the early retirement routes for women and long-term insured had a *de facto* closure date of 2016, due to their availability to particular cohorts (OECD, 2005f).

Deferred retirement

Individuals could increase their pension amount by 0.5% for every month they worked after the age of 65 throughout the period under investigation.

Labour market policies

Early exit

In Germany prior to 1986, the '59er' provision meant if an individual was unemployed at 59, they could claim unemployment benefit for one year before accessing an early pension. The 1984 Early Retirement Act (*Vorruhestandsgesetz*) was introduced to reduce access to the expensive '59er' scheme by regulating agreements between Trade Unions and employers for compensation for those who exited voluntarily. As a result, in the 1980s, it was common practice in Germany to make an individual redundant from the age of 58 (or younger) with their agreement, thereby circumventing national legislation. The employer would pay the difference between the previous wage and unemployment benefit through a compensation payment referred to as *ratierte Abfindungen* (OECD, 2005f). This Act was in place from 1984 to 1988 and allowed workers to retire at 58, providing their employer paid them an income equal to a minimum of 65% of their final net wages for between two (for women who could then retire early at 60) and five years (for men who could retire early at 63) until the individual reached the state pension age (Mandin, 2004). This stipulation was designed to prevent the older individual moving to disability benefits and thus reduced the cost for the state (Teipen and Kohli, 2004). Due to the cost for employers, 165,000 people used this scheme between 1984 and 1988 when it expired. The government refrained from narrowing the options for early exit during the late 1980s; the Pensions Reform Act 1992 made moves to curb early exit but it was originally intended that its effects would not be seen until 2001 (Mandin, 2004).

However, despite concerns regarding the cost of the '59er' provision, in 1986, the ruling on early pension access was altered under Regulation 428 SGBIII which decreed that those over the age of 58 would no longer be offered assistance in job search or be required to look for work. This lower age limit, when combined with unemployment benefit receipt, meant the individual could effectively exit at 57.4 years of age (thus becoming the '57er' provision) (Guillemard, 1993). These individuals could then exit early at 60 under the early retirement route for the unemployed if they had made the requisite contributions (OECD, 2005f; Walker, 1993). As a result, in 1988 only 18.3% of retirees exited at the normal age of 65 and 18.4% exited at 63 under flexible arrangements. Thus only 36.7% went onto pensions straight from work; the rest utilised other forms of social protection and withdrew from the labour market earlier. Of these, 38.1% entered the Invalidity Insurance Scheme, 12.5% were declared unfit for work and 12.6% were unemployed; these groups could get a pension at 60 but none went directly from work to retirement (Walker, 1993).

With regard to early exit through conventional unemployment benefits, these were divided into unemployment insurance and assistance, both

of which had contribution requirements. The former was more generous, and the durations were dependent on age and contribution record
whilst unemployment assistance was unlimited. The maximum duration an individual aged 54 could receive was 832 days, providing they
had made 1,920 contribution days; for an individual the same age,
this was reduced to 728 if they had completed 1,680 contribution days.
For those under the age of 42, the maximum benefit duration was
312 insurance days.

In 1992, unemployment insurance access was lengthened so as to
benefit older long-term unemployed waiting to reach state pension
age. Subsequently disability benefits were closed off whilst unemployment benefit criteria were widened, thus "[r]efocusing the movement
towards early exit from invalidity insurance is also for the Federal
German State a way of redistributing the cost of the process so as to
lighten somewhat the heavy burden on state funds" (Guillemard,
1993: 41).

By the late 1990s, the consensus around early exit had been abandoned in the face of demographic ageing and potential labour shortages.
An indication of this was the passage in 1992[8] of a law prohibiting
obligatory early exit in employment contracts (Taylor, 1998). As part
of the 1999 reforms, the unemployment insurance duration had also
been modified to provide individuals over the age of 52 with a maximum of 26 months of insurance if they had made 57 months of contributions within the previous seven years; for those aged 57 and over,
the insurance benefit could be received for longer (32 months) if they
had made 64 months of contributions in the same period. For the rest
of the population under 45 years of age, the maximum duration was
12 months (OECD, 2001).

The *Hartz I–IV* reforms of 2003–5 altered the unemployment benefit
system (Frericks and Maier, 2008). The maximum period of unemployment insurance receipt (*Arbeitslosenversicherung*) was reduced to 18 months
for those over 55 after which the Basic Resources for Jobseekers (*Grundsicherung für Arbeitsuchende*) would have been the only option. In addition, from 2006 those over the age of 58.5 would have been required
to engage in job search. With regard to job search and Unemployment
Benefits I, individuals were expected to take up any job that corresponds
to their working capacities, irrespective of their personal feelings or preferences. The maximum commuting time is 2½ hours a day for an eight-
hour job or two hours a day for a six-hour job, and the unemployment
agency can request relocation after four months of unemployment (unless
there are important family reasons). For Unemployment Benefit II, the

criteria are different and an individual has to accept any job they are physically or mentally capable of doing, regardless of salary or qualification level (Tergeist and Grubb, 2006).

Active labour market policies

In light of the predictions around population ageing, a targeted employment promotion law in 1995 provided wage subsidies for companies who employed those over 50 who had been unemployed for one and half years with 70–75% of standard wages for up to eight years. Measures were introduced that specifically targeted older women and those from minority ethnic groups (Taylor, 1998). However, Frerichs and Naegele (1998) argue these corrective measures came into force too late as they focus on the long-term unemployed. In addition, an integration subsidy (*Eingliederungszuschüsse*) equal to 50% of the wages was provided to employers who employed individuals 55 years and over who had previously been unemployed for six months. The subsidy would be paid for three years and was available from 1998 to 2003 (OECD, 2000, 2005f).

As part of the 1999 reforms, to combat the problem of long-term unemployment, the aforementioned integration subsidy was made available after a period of six months as opposed to 12 in 1999 (OECD, 2000). In addition, employers were no longer obliged to retain this worker when the subsidy expired. This, it was argued, would remove barriers to employing another unemployed person (Federal Republic of Germany, 1999). This year also saw the definition of 'older employees' reduced from 55 to 50. However, for those over the age of 55 the length of support time was increased from three to five years. In this year, increased training opportunities were proposed subject to the consent of the social partners and working groups within the *Bündnis für Arbeit, Ausbildung und Wettbewerbsfähigkeit* (Alliance for Jobs, Training and Competitiveness) (Federal Republic of Germany, 2000). The Alliance for Jobs was reformed in 2001 to allow those over 50 to receive Integration Subsidies (Committee of the Regions, 2003). The Job-AQTIV Act launched on 1 January 2002 meant that the unemployed individual, in conjunction with the employment office, was required to create a profile and a plan to improve employability. This measure was designed to prevent people drifting into long-term unemployment.

In 2002, a new initiative was launched called 'A New Quality of Work' (*Neue Qualität der Arbeit*) (Federal Republic of Germany, 2002). Also, the Hartz Commission of 2002 created exemptions for employers who hired unemployed individuals over the age of 55 and lowered the age limit for fixed-term employment from 58 to 52 years to improve the prospect of reintegration (Taylor, 2005).

2003 saw the reform of 'mini-jobs', designed to increase low wage employment. Mini-jobs had been first reformed in 1999 when they ceased to exempt employees from social security contributions. Before this point, mini-jobs had wages below a threshold and a maximum of 15 hours per week. Workers were also able to hold several mini-jobs at the same time. From 1999, employers had to pay 22% of social security contributions for each mini-job and workers could choose between paying a 20% flat-rate tax on earnings or to pay tax according to the income. As of 2003, the hours' restriction on mini-jobs was abolished and the earnings threshold increased. In addition, the tax rules were modified so as to encourage low paid work. From this point, contributions would begin at 4% and increase linearly to the standard rate of 21%. The mini-job scheme would be a key part of the German welfare-to-work strategy (Bertelsmann-Stiftung Foundation, 2003c). In this year the integration subsidy (*Eingliederungszuschüsse*) was disbanded (OECD, 2005f).

As part of the Modern Labour Market Services Act of 2003, incentives were introduced for the long-term unemployed to enter the labour market. To encourage the take-up of employment at a lower age, the Wage Guarantee was available from 2003 for those unemployed and over 50. This scheme provided a subsidy of upon re-entry into the labour market at 50% of the difference between the new and previous wages. The individual had to have 180 hours left of unemployment benefit, which was then the period for which they received the Wage Guarantee. This scheme was closed to new entrants at the end of 2005 (OECD, 2006). In addition, employers who recruited individuals over the age of 55 were exempted from social insurance contributions (Committee of the Regions, 2003; Federal Republic of Germany, 2003).

The Employment Promotion Act was modified in 2004 to only cover those businesses with ten or more employees as opposed to the previous five in order to provide relief for smaller enterprises. Older workers with long service records with a particular firm were in theory protected from redundancies in that their employer was obliged to cover this unemployment benefit (the *Erstattungspflicht* ruling). This criterion was made more stringent in 2004. In addition, the range of 'social criteria' employers could use in cases of collective dismissal had been limited to disability, age, job tenure and whether the individual had dependents in their household. The employee was also obliged to take up severance pay equivalent to half a month's salary instead of challenging the dismissal in court (OECD, 2005f). In terms of specific ageism legislation, Article 36 of the Social Security Code (*Sozialgesetz-buch*, SGB III) ruled there could only be age limits in job adverts if they could be justified and the Law on Labour Relations at the Workplace (*Betriebsverfassngsgesetz*, BVG) stated that employers and works councils could not discriminate against older workers. The latter could instigate measure to promote older workers' employment (Articles 75, 80, 96) (Taylor, 2005).

The *Hartz* reforms of 2004 aimed to accomplish three things: 1) to improve employment services and policies; 2) to activate unemployed individuals and, 3) to increase demand for employment by deregulating the labour market (Jacobi and Kluve, 2006). As part of this reform, temporary work and subsidies for those over the age of 52 were introduced (OECD, 2005f). This, it was hoped, would encourage the recruitment of older individuals (OECD, 2005f). To this end, subsidies were also introduced to cover 50% of the wages of newly recruited workers over the age of 50 for up to three years. In addition, those over 50 who took up low paid work could receive 50% of the difference between their new wage and previous income. This amount was payable until the point where benefit entitlement ended. In addition, the Hartz reform also made employers who recruited individuals over the age of 55 exempt from unemployment insurance and reduced the age limit for fixed-term employment from 58 to 52 (OECD, 2005f; Frericks and Maier, 2008; Jacobi and Kluve, 2006).

Further ALMPs were introduced from the mid-2000s, including 'Employment Pacts for Older People in the Regions', launched in 2005 and ran until 2007, which made €250 million available for the purpose of the activation of older workers. This programme was then re-launched in 2008 to run for two years with a budget of €275 million. In addition, there was the New Quality of Work Initiative which aimed to look into extending working lives through workplace design and conditions, as well as promote skills and workability. The Initiative also raised awareness among employers about demographic change and modern workplace design. May 2007 saw the launch of the Act on Improving the Employment Prospects of Older People, which contained a number of elements. First, it implemented integration subsidies for individuals over the age of 50 (Section 421f SGB III). Employers could then receive 30–50% of the older workers' wages for between 12 and 36 months. Second, the Act also stipulates that individuals over the age of 50 who left unemployment benefit for work which is of lower pay than their previous employment would receive a subsidy to partially cover the difference, 50% for the first year and 30% for the following year (Section 421j SGB III). Also, in firms with less than 250 employees, workers over the age of 45 would receive additional payments for vocational training (Section 417(1) SGB III). Finally, individuals over the age of 52 and were unemployed for four months, have received *Transferkurzarbeitergeld* (transfer short-term allowance) or have participated in publicly funded employment measures under the SBG II or SGB III, were permitted to move to employment under fixed-term contracts with a duration of up to five years without specific reasoning (Federal Republic of Germany, 2008).

Part-time work/pension arrangements

In an attempt to move away from early retirement towards gradual retirement, the aforementioned Early Retirement Act was not renewed in 1988[9] but replaced in 1989 with the German Old Age Part-Time Employment Act. This legislation established a rule whereby those over 58 who had been insured for three of the previous five years could reduce their working time by up to 50%. This reduction was flexible and allowed the individual to 'block' time over the agreement. This was modified in 1992 to allow the individual to receive some of the full-time benefit (OECD, 2005f).

In 1996, the part-time employment scheme for older workers was revised and re-launched. Previously, the main deterrent for entering into part-time work as opposed to an early pension was the loss of income. The Act on Part-time Work in Old Age therefore ameliorated this risk by covering the gap between wages for full-time and part-time work (Federal Republic of Germany, 1998) and encouraging part-time employment for older workers as opposed to the complete premature exit from the labour market. Those who engaged in this scheme would receive a minimum of 70% of their previous earnings plus at least 90% of their pension fund contributions from their employer (Federal Republic of Germany, 1998). The state would then reimburse the employer, provided an unemployed individual was recruited to make up the working hours (OECD, 1997; Federal Republic of Germany, 1998; Lafoucrière, 2002; Teipen and Kohli, 2004). Individuals could access this benefit from the age of 60.

On 1st January 2000 the Law on Development of Part-Time Work for older workers was activated. This extended the previous policy to companies with less than 50 employees to provide a subsidy if an older employee moved into part-time work and the remaining hours were made up by an individual from the unemployment register. The supplement would begin at 0.5% of gross wages, rising to 4% in 2008. In this year, 20 billion DM was allocated for subsidies until 2008 (Bertelsmann-Stiftung Foundation, 2000a). In addition, the coverage of the existing part-time pension was extended to include those already engaged in part-time work so as to avoid indirect discrimination along gender lines (OECD, 2005f).

As part of the 2004 reforms, individuals could reduce their pension to either $\frac{1}{3}$, $\frac{2}{3}$ or $\frac{1}{2}$ and continue to work for a proportional period of time (Federal Republic of Germany, 2004). From 2006–8, the age threshold for the part-time pension was increasing from 60 to 63 (Social Security Administration, 2006).

5.4 Summary and impact at the individual level

In terms of the common policy ground shared by the Netherlands, Finland and Germany, in the early 1990s, all three had comparatively extensive ranges of early exit and retirement routes, which is perhaps reflected by their low employment rates for older individuals during that decade. Indeed, these policies were at the time seen as an effective means of reducing youth unemployment. All nations began to move away from this policy approach from the late 1990s/early 2000s onwards with the retrenchment of decommodification options whilst at the same time introducing ALMPs including subsidies and anti-ageism legislation in the Netherlands, lifelong learning opportunities in Finland and increased conditionality and work compulsion in Germany.

At the micro-level, these nations presented mixed decommodification approaches in 1995 when the data from the early exit and early retirement policies are addressed separately. In Germany, the 50-plus model biographies (see Section 3.6) would only have been able to potentially access the early exit routes due to the age thresholds for the early retirement schemes, as well as the partial pension schemes. Over the entire period for *Laurent* (55 with 35 years of employment contributions) and *Jean* (55 with 35 years of employment contributions) the unemployment benefit duration was contingent on age and contribution record, thus allowing *Laurent* to exit early. In addition, there was the option of subsidised employment from the age of 50. The data regarding *Sasha* (63 with 35 years of insurance contributions) and *Jude* (again 63 with a disjointed work history) in Germany reveals that when age is no longer a factor, the ability to be decommodified was bestowed on a mix of rights and desert-basis. There was a stronger focus on the latter in 1995 in terms of the early retirement options specifically for women, for those with a reduced working capacity, the unemployment pension and long insurance duration scheme all required a certain level of contributions (35 years with the exception of the scheme for women which was 15 years). *Jude* however would have been exempted from the job search requirement of unemployment benefit receipt for the entire ten year period. Thus though there was a rights-element, desert-based routes were more prevalent. However, the policy picture had changed dramatically by 2001 for *Sasha* as there had been significant retrenchment of the early retirement including penalties resulting in a reduced pension amount by between 18–23.4% (due to the different age thresholds), depending upon the route accessed by *Sasha*. The age thresholds had also been raised to 63 years. Thus desert was the key principle of decommodification, and by 2001 this came at the price of a reduced pension. At the same time, ALMPs were expanded.

Further retrenchment of decommodifying policies was undertaken by 2005, though they did not apply to *Sasha* and *Jude* due to the year of their birth. A new system had been introduced for younger birth cohorts which provided limited decommodification potential. Thus it would seem that the retrenchment of the desert-focused decommodification options had gone as far as was politically viable in terms of increases to the age thresholds and the introduction of penalties over the period of 1996 to 2001; ultimately these 'deserved' schemes could not be closed entirely to those who had already contributed significantly but they could be barred to new potential claimants, thereby creating a *de facto* abolition date. In addition, the labour market was made a more attractive and viable option through the creation of ALMPs, though *Sasha* and *Jude* would still have been able to exit via unemployment benefits until they reached the state pension age. The *Hartz* reforms had however made these less generous, with a maximum period of Unemployment Insurance (*Arbeitslosenversicherung*) limited to 18 months after which the Basic Resources for Jobseekers (*Grundsicherung für Arbeitsuchende*) would have been the only option. In addition, both *Jude* and *Sasha* would no longer have been exempted from actively seeking work as of 2006.

When the data from the 50-plus and 60-plus model biographies are combined in the case of the Netherlands, the decommodification approach represents a mix of rights and desert-foci. Thus in 1995, had their working capacity been reduced, all the model biographies could have retired early with no additional requirements. However, in terms of the unemployment benefit durations, these were dependent on age and contributions, with an additional period for those over the age of 57.5. *Laurent* would therefore have been able to access an extended duration due to their contributions whilst *Jean* would not; both were too young for the additional amount from those over the age of 57.5. For *Sasha* and *Jude,* their age meant they could have received unemployment benefits until the state pension age and would not have been required to engage in job search, as well as being eligible for the partial pension scheme. Though the routes were limited, retrenchment was undertaken and by 2010 all individuals would have been required to engage in job search if unemployed and those already receiving the early retirement scheme for reduced working capacity (WAO) were required to be reassessed (new applicants would instead be eligible for the WIA which had an increased focus on re-employment). At the same time, ALMPs were introduced to increase older individuals' labour market prospects.

Finland too demonstrated a mix of desert- and rights-foci in the early 1990s when the data from the four model biographies was combined. In 1995 *Laurent* and *Jean* (providing the former met the contribution requirement for unemployment allowance of 26 weeks employment in the previous two years, otherwise they would have received a means-tested labour market subsidy from the state for an unlimited period) could have exited early via the 'unemployment tunnel' for which the entry requirement was age, and the former could also have accessed the early retirement scheme for those employed in the private sector with a reduction of 6% per year to their final pension amount (individual early pension (*Yksilöllinen varhaiseläke*)). However, with regard to the model biographies over the age of 60, Finland exemplified a mix of rights- and desert-focused decommodification approach in that *Sasha* had more choice than *Jude* due to their contribution record, but the latter was also able to access some early retirement schemes (*Sasha* could have accessed the unemployment pension (*Työttömyyseläke*) and individual early pension (*Yksilöllinen varhaiseläke*), again with a 6% reduction whilst both could have utilised the early old age pension (*Varhennettua Vanhuuseläkettä*) with a 6% reduction and the farmer's pension). Both *Sasha* and *Jude* were able to combine work and pension receipt.

Retrenchment of decommodification options in Finland was undertaken comparatively early from 1997 onwards. Due to the reform of the individual early pension (*Yksilöllinen varhaiseläke*), *Laurent* was no longer eligible as the age threshold had been raised to 60 by 2000. In terms of the schemes available to the 60+ model biographies, both *Sasha* and *Jude* were still eligible for the farmers pension and the old age pension (*Varhennettua Vanhuuseläkettä*), and *Sasha* could also still access the unemployment pension (*Työttömyyseläke*) and individual early pension (*Yksilöllinen varhaiseläke*). At the same time, the reduction per annum was lowered for the individual early pension and the early old age pension to 4.8% which could have in turn potentially decreased the disincentive effects for claimants. A reduction was also added to the unemployment pension (*Työttömyyseläke*) of 4% per year prior to the state pension age in 2000. However, within five years, these choices had been significantly curtailed with the closure of the individual early pension (*Yksilöllinen varhaiseläke*) to new entrants and the unemployment tunnel age threshold had been raised to 57. In 2005, the option of flexible retirement was introduced with accrual rates which incentivised working at older ages; thus *Laurent* and *Sasha* may have been better placed to take advantage of this flexibility as their pension amount would reflect their contribution years whilst *Jean* and *Jude* may have found further reductions too significant. By 2010, the unemployment pension was also closed off to those born after 1950 and thus *Sasha* would still have been eligible.

These three nations all made progress beyond the Stockholm target between 2001 and 2010 and all made significant retrenchment of their early exit and retirement routes. These nations had in common decommodification options which exemplified a mix of rights and desert as their eligibility focus. Not only did these nations curtail the universal, rights-focused decommodification options, they also restricted access to the opportunities for exit which required certain levels of contributions, either by raising these, increasing age thresholds or reducing the final pension amounts of those choosing these routes. Thus the expectation that individuals should remain active in the labour market in older age was apparent, even where previously early exit or retirement could be 'deserved'. Finland represents an interesting case as on the one hand the new system of pension accrual could be seen to penalise those unable to participate in the labour market. However, by increasing the percentage accrued by those aged 63–67, this could potentially allow older people whose previous employment had been interrupted by, for example, the provision of care, to make up for these lost contribution years (the accrual for those aged 63–67 was three times that of those aged 18 to 52).

6
Group III: Below Stockholm but Approaching Fast

Of those nations that were still below the Stockholm target, some had made more progress than others. Austria, Belgium and Luxembourg were still under this 50% goal for 55–64 year olds' labour market participation and the latter two also remained the lowest among all EU15 nations for 55–64 year olds' employment rates. However, in terms of the progress they had made, it must be acknowledged that these nations had the most ground to make up, having the lowest employment rates in 2001. Thus the increases they have made need to be considered and all three have surpassed the average rise for EU15 nations in total (9.6%).

6.1   Austria

The employment rate for older workers in Austria was low until 2005 when it began to rise, before which point it had largely remained below 30%. By 2010, it had still not reached 50%, but had increased by over 10% in five years. Thus since 2001, the employment rate had risen by 13.5% from 28.9% to 42.4%, which is significantly higher than the average increase of all EU15 nations (9.6%), but is still below the EU15 average of 48.4 (Eurostat, 2011). Thus Austria has made significant moves towards the Stockholm target, and the reforms of 2005 which created a new system of exit for those under 50 could help progress in the future (see *Early retirement*).

Pension policies
Pension principles and the state pension age
Austria's public pension scheme rests on a defined-benefit principle with an income-tested supplement for those below a certain threshold (Whitehouse, 2007). Established as part of the General Social Insurance

Act (*Allgemeines Sozialversicherungsgesetz*, ASVG) in 1955, this mandatory pay-as-you-go pension system is for all workers with the exception of the self-employed and those in very low-paid work.

In the early 1990s, in order to receive a full standard old age pension (*Regelaltersrente*), individuals had to have worked for 40 years and the minimum period of membership operated on a principle of 'eternal eligibility' (*ewige Anwartschaft*) which applied to those who had contributed for either 180 of the previous 360 months or 300 contributions months in total, including 'fictitious qualifying periods' (*Ersatzzeiten*) such as military service, receipt of maternity allowance (*Wochengeld*), unemployment benefits (*Arbeitslosengeld*) and sickness benefits (*Krankengeld*). The state pension age was 65 for men and 60 for women in the years leading up to 1995 and will be harmonised between 2024 and 2033 as part of the 1992 Constitutional Act on Unequal Retirement Age (Schulze and Schludi, 2007).

The *Pensionssicherungsreform* of 2003 came into effect on 1st January 2004. As a result of this reform, the assessment base for the full standard old age pension (*Regelaltersrente*) will be gradually extended from 15 years to 40 by 2028. Also, the replacement rate was reduced with the increment points for each year of pension contributions falling from 2% to 1.78% until 2009. As a result of this change, 40 years of contributions would result in a payment of 71.2% of the assessment base as opposed to the former 80%. The amount a pension could be reduced by as a result of the reform was limited to 10%. The assessment base is being gradually modified to take into account an increase from the best 15 to the best 40 years until 2028. The amount of pension credit awarded per child was increased from 18 months to 24 (OECD, 2011).

Large-scale reform of the pension and early exit routes was also undertaken in 2004 with the *Pensionsharmonisierungsgesetz*, which came into force on 1st January 2005. The OECD (2005a) argues this reform instigated the creation of a harmonised system based on individualised pension accounts. Parallel accounts were set up to allow the calculation to be weighted according to the average number of credits in the system pre- and post-harmonisation (i.e. in accordance with the reforms of 2003 and the subsequent changes in 2004/5). For persons who had not yet reached the age of 50 by 1 January 2005 and had not made insurance contributions, the 'eternal eligibility' (*ewige Anwartschaft*) was altered to 180 months of which 84 could not be *Ersatzzeiten*. For those over the age of 50 on January 1st 2005, the previous 'eternal eligibility' of 180 insurance months within the last 360 calendar months or 300 contributions months in total still applied. In addition, new periods of absence from the labour market (*Ersatzzeiten*) were included as qualifying for pension credits as of 2005, such as parental leave, sickness and military service (OECD, 2005a).

Early retirement

Since 1960, women aged 55 and men aged 60 could retire if they had made 35 or 37.5 years of contributions respectively under the *vorzeitige Alterspension bei langer Versicherungsdauer* (Special Regulation for Very Long Insurance) scheme. With the rise of unemployment in the 1970s, 1979 saw the introduction of pre-retirement schemes for all older individuals who were unemployed and had made 20 years of contributions, allowing them to exit five years prior to the pension age (*vorzeitige Alterspension bei Arbeitslosigkeit*) (Vogt, 2006). In addition, disability pensions for unemployed miners (*Sonderunterstützung Bergbau*) and those engaged in strenuous night work (*Sonderruhegeld*) were also introduced in this period (1981 and 1973 respectively) (Committee of the Regions, 2003; OECD, 2005a). With regard to the latter, the individual had to have been engaged in heavy night work for 20 years or 15 years within the previous 30. This scheme allowed women to exit at 52 and men at 57. The early pension for miners stipulated that they had to be aged 51 and over and had been employed in mining industry for ten years prior to unemployment. These schemes had relatively few beneficiaries with the mining scheme being received by 1,900 people and the options for those in strenuous night work available to 1,200 people in 2003 (OECD, 2005a). As part of the 1992 Constitutional Act on Unequal Retirement Age, the early retirement age thresholds for women were being raised from 55 to 60 between 2019 and 2028 (Schulze and Schludi, 2007).

Also an early pension was in place since 1993 for those with a reduced working capacity (*Vorzeitige Alterspension aufgrund geminderter Arbeitsfähigkeit*) over the age of 55 (this was separate from the disability pension for which there was no age threshold). The individual either had to have achieved the eternal qualifying period (240 contribution months) or 180 insurance months within the previous 360, though this requirement was waived if the individual's capacity had been reduced by an accident at work. This route provided exit for those whose working capacity was reduced so as to make the engagement in the employment they had undertaken for the previous 15 years impossible (OECD, 2005a).

In 1997 the age threshold of the pension for those with a reduced working capacity (*Vorzeitige Alterspension aufgrund geminderter Arbeitsfähigkeit*) was raised to 57 for men (OECD, 2005a, 2006). This and some of the other early pension routes (for those with very long insurance records (*vorzeitige Alterspension bei langer Versicherungsdauer*) and the long-term unemployed (*vorzeitige Alterspension bei Arbeitslosigkeit*)) were also modified in 1997 to include a bonus-malus element, resulting in a reduction or increase of 2% for those who exited the labour market.

The Pact for Older Persons (*Pakt für ältere ArbeitnehmerInnen*) of 1999 came into place in 2000 (OECD, 2005a, 2006). As part of this Pact, the early retirement age thresholds for the schemes for long insurance durations (*vorzeitige Alterspension bei langer Versicherungsdauer*) and the long-term unemployed (*vorzeitige Alterspension bei Arbeitslosigkeit*) were raised by two months at the start of each quarter from October 2000 to achieve an increase of 18 months by October 2002, thus from 55 to 56.5 and 60 to 61.5 for women and men respectively (OECD, 2005a, 2006). In order to increase the actual retirement age of the population, the reductions for retiring prior to the state pension age were increased from 2% to 3% and the bonus for deferment increased to 4% (Schulze and Schludi, 2007). At the same time however a route was created for those with very long insurance records (*Hacklerregelung für Langzeit-versicherte*, 45 years for men and 40 for women) who were born before 1949 if male and 1954 if female. Thus this route would have a *de facto* expiration date of 2010 and allowed men and women to retire five years prior to state pension age. From 2000, those men born in 1940 and women in 1945 (increasing by one year every year after this point) received credits which could enable access to unemployment-based early retirement without having to be eligible for unemployment assistance (Republic of Austria: Federal Ministry for Economic Affairs and Labour, 2001; Republic of Austria: Federal Ministry for Economic Affairs and Labour, 2002).

The pension for those with a reduced working capacity (*Vorzeitige Alterspension aufgrund geminderter Arbeitsfähigkeit*) was also abolished in 2000 and the criteria for the disability pension (which was open to all ages) were relaxed for those over the age 57 with regard to individual assessment of activity and therefore accommodated those who would have been eligible for the former route (OECD, 2005a, 2006). In order to receive the disability pension for those with a loss of earning capacity of 50% compared to an insured person with the same level of education, the individual also had to have made 60 months of contributions in the previous ten years, plus an extra month for each month from the age of 50, in addition for 300 insurance months and 180 months of contributions (Social Security Administration, 2004; Social Security Administration, 2010).

As part of the *Pensionssicherungsreform* of 2003 which came into effect in 2004, the reductions for those retiring early were raised to 4.2% per year. In addition, the *vorzeitige Alterspension bei Langer Versicherungsdauer* (Special Regulation for Very Long Insurance) scheme was to be phased out from 2004 to 2017. The early retirement route for long-term unemployed individuals (*vorzeitige Alterspension bei Arbeitslosigkeit*) was also closed off as part of the *Pensionssicherungsreform*.

As part of the large-scale reform of the pension and early exit routes which became active on 1st January 2005, for those under 50 in 2005, a 'retirement corridor' (*Korridorpension*) for men aged between 62 to 68 was created, with 65 acting as the reference pension age. Women too would be subject to this scheme as of 2028 when their pension age will be harmonised with men; until then, retirement can be accessed at 60. Under the retirement corridor scheme, men taking their pension before the legal retirement age of 65 would experience a reduction to their pension payment of 4.2% per year with a maximum reduction of 15%. Conversely, those who delayed retirement up to 68 would receive a bonus at the same rate (Social Security Administration, 2004).

With regard to existing early retirement arrangements available to those older than 50 in 2005, a number of changes were undertaken. From the second half of 2004 the early retirement age was increased by two months for every quarter, reducing to one month after 2004 (OECD, 2005a). In terms of the early retirement routes for specific groups, as of July 2004, the early retirement scheme for those with extensive insurance contributions (*vorzeitige Alterspension bei langer Versicherungsdauer* – Special Regulation for Very Long Insurance) was gradually being phased out with an increase of one year to the threshold until it will reach the state pension age in 2017 (Republic of Austria: Federal Ministry for Economic Affairs and Labour, 2004).

From 2007, there was an early pension (*Schwerarbeiterragelung*) for male employees engaged in arduous work which allowed individuals aged 60 with 45 years of insurance coverage, of which at least 120 months were in physically demanding jobs in the previous 240 months. For these individuals, exit could occur three months earlier for every year over a threshold of 15 years in strenuous employment with a limit of age 60 and a smaller reduction of 2.1% per year prior to the state pension age. This early pension is not available to women due to the lower state pension age, but a new scheme for women in arduous labour will come into effect in 2028 once the early retirement ages have been harmonised (Social Security Administration, 2006; OECD, 2005a).

By 2010, the early pension age threshold for men was 62 and for women it was 57.5. This age threshold is gradually increasing by one month per quarter to reach the state pension ages of 65 for men and 60 for women by 2017 for those with between 420 and 450 months of insurance coverage (*vorzeitige Alterspension bei langer Versicherungsdauer* – Special Regulation for Very Long Insurance). As it was only available for those born before 1949 if male and 1954 if female, the early pension for long-term insured (*Hacklerregelung für Langzeitversicherte*) by this year was effectively phased out (Social Security Administration, 2010). The disability pensions for unemployed miners (*Sonderunterstützung Bergbau*) and those engaged in strenuous night work (*Sonderruhegeld*) were unreformed (OECD, 2005a). For those who were under 50 in 2005, the 'retirement corridor' (*Korridorpension*) for men aged between 62 to 68 was available, with a reduction of 4.2% per year prior to 65 with a maximum reduction of 15%. This scheme will also be available to women as of 2028 when their pension age will be harmonised with men (Social Security Administration, 2004).

Deferred retirement

Individuals deferring their pension receipt beyond the age of 60 for women and 65 for men would in 1995 receive an increase of 2% between the ages of 61 to 65, 3% from 66 to 70 and 5% from the age of 70 onwards. Individuals deferring their pensions for a maximum of five years would receive a lump-sum bonus from 1998. By 2000, the individual would receive a 4% increase for every year they worked after the state pension age. This was increased to 4.2% per year up to 12.6% in total in 2003 as part of the *Pensionssicherungssreform*. As part of the retirement corridor scheme introduced in 2005 for those under 50 in that year, individuals deferring their pension up to the age of 68 could also increase their pension by 4.2% per year with a maximum increase of 12.6% (OECD, 2005a).

Labour market policies

Early exit

In the 1990s an early exit route was introduced, allowing individuals to leave the labour market prior to the state pension age via the unemployment benefit system. Since 1993 those who had contributed 180 months to the public pension scheme and had been unemployed for a year could receive the pre-retirement benefit (*allegmeine Sonderunterstützung*). This benefit provided women over the age of 54, men over 59 (with

180 insurance months) with a 'pension' for one year until they reached the early retirement age. There were no additional unemployment benefits or supplements available on the grounds of age. In the early 1990s, all older unemployed individuals were exempt from the job search requirement of benefit receipt.

The policy approach to older individuals began to shift and the mid-1990s is recognised as being a period of prudence with the introduction of the Austerity Programmes in 1995–1996 to curb public spending in line with Maastricht criteria. As of 1996, older individuals were no longer exempt from job search which had served to remove them from unemployment figures (OECD, 2006). In addition, the pre-retirement benefit (*allegmeine Sonderunterstützung*) was technically abolished in 1996; yet due to transitionary arrangements, continued to run until 2004 (OECD, 2000).

However, in spite of the shift away from early exit and retirement, in 2001 the maximum period for receipt of unemployment benefit was extended to 78 weeks for those over 50 who had contributed nine out of the previous 15 years (compared to 15 out of the past 25 years for the rest of the population) (Republic of Austria: Federal Ministry for Economic Affairs and Labour, 2001).

The early retirement route for long-term unemployed individuals (*vorzeitige Alterspension bei Arbeitslosigkeit*) was closed off in 2004 following the *Pensionssicherungssreform* of 2003, as discussed above in *Early retirement*. In addition, the pre-retirement benefit (*allegmeine Sonderunterstützung*) came to a close in the same year. Thus older unemployed individuals from this point onwards were subject to the same conditionality as their younger counterparts, though they could receive additional advice tailored to their age group as unemployment benefits were no longer a *de facto* early exit route and the *de jure* schemes had also been closed off (OECD, 2005a). However, to replace the route for the long-term unemployed, a transitionary benefit (*Ubergangsgeld*) was introduced in 2003 to cover those who would have been eligible until 2006 (OECD, 2005a).

Active labour market policies

In the mid-1990s, a period recognised as representing a sea-change in Austria's policy approach to older workers, a comprehensive range of targeted active labour market policies were launched. In 1996, a 'bonus-malus' scheme (OECD, 2005a) was introduced whereby the unemployment contributions for employees between the ages of 50 and 55 were halved and completely eliminated for those over the age of 55. Accordingly, a penalty was applied to employers who made older individuals redundant on a sliding-scale basis (OECD, 2005a). In 1999, the AMS developed non-profit and private temporary work agencies for the 50-plus as well as the implementation of job coaching for older workers in the early stages of unemployment.

Within the 1999 Pact for Older Workers, there was also a shift towards more active measures in terms of employment policy including flexible working arrangements for older employees and more targeted measures from the AMS (including the removal of age limits for training). Under section 38a of the Austrian Public Employment Service Act (*Arbeitsmarkt-servicegesetz*, AMSG), from 2000 onwards individuals who met the eligibility criteria for early retirement were also entitled to a place on an employ-ment project as a means to promote continued labour market parti-cipation (Federal Ministry for Labour et al., 1999). Education benefits (*Weiterbildungsgeld*) were increased to match unemployment benefits. However, from January 2000 under the Pact, the minimum period for taking educational leave had been reduced from six to three months for older workers (Federal Ministry for Labour et al., 1999).

The Pact for Older Persons also created measures to retain and re-employ older individuals. The bonus-malus system with regard to employer prac-tices was improved in October of 2000 to exempt those who recruited the 50-plus from any unemployment insurance contributions. On the 'malus' side, the basic amount for terminating the employment of a 50-plus employee after ten years of service was doubled (Republic of Austria: Federal Ministry for Economic Affairs and Labour, 2001; Committee of the Region, 2003). Age discrimination received special attention in 2000 with modi-fications made to existing legislation to protect older workers from unfair dismissals. As of 2000, companies were obliged to inform the Labour Market Service (AMS) if they made redundant more than five 50-plus workers regardless of the size of the company. In addition, older workers who had become recently employed could appeal against the termination of their contract if it was for 'socially unjustifiable' reasons had they been re-employed for two years (pre-2003, this had only been six months) (OECD, 2005a: 13).

As part of the raft of policies introduced in 2001, the Labour Market Promotion Act (*Arbeitsmarktfördesrungsgetz*, AMFG) was altered to include an 'early warning system' so as to prevent redundancies through subsidies, flexible working-time practices, training for older workers and part-time benefits (Republic of Austria: Federal Ministry for Economic Affairs and Labour, 2001). These preventative measures were introduced to assist older individuals before they drifted into long-term unemployment.

Further additions were made to Article 45 of the Labour Market Promotion Act to state that the AMS was to be informed of any redundancies of those 50-plus by the day notice was given to the employees (providing they had been with their employer for a minimum of six months). Following this, the AMS would provide guidance in an attempt to find these individuals work with their current employer or in a new job (Republic of Austria: Federal Ministry for Economic Affairs and Labour, 2002). In 2002 severance pay was reformed to become *Abfertigung Neu* which implemented a more universal system, based on contributions. This modification meant that all employees were covered as of one month, a sharp increase of the previous three year threshold. Entitlement to this protection increased over time with employers contributing 1.53% of monthly wage.

The *Pensionssicherungsreform* of June 2003 included measures to tackle the problems associated with demographic ageing and pension sustainability. Special measures were introduced to prevent older people entering long-term unemployment. Those over 50 were offered participation in training within 12 weeks of becoming unemployed (Republic of Austria: Federal Ministry for Economics and Labour, 2003). *Aktion 56/58* removed employers' unemployment contributions for women and men over the ages of 56 and 58 respectively. When the employee reached 60, the employers were also exempt from the surcharge levied under the Act on Wage Compensation from the Insolvency Contingency Fund (IESG) as well as contributions to work accident insurance and to the Family Burdens Equalisation Fund (Republic of Austria: Federal Ministry for Economics and Labour, 2003). Linked to this, from 2003, anti-ageism legislation was modified by applying protection in cases of dismissal to only those 50-plus employees who had been in that particular company for two years (Republic of Austria: Federal Ministry for Economics and Labour, 2004). However, employers could enforce mandatory retirement on workers who had reached state pension age (OECD, 2006).

In 2004 the unemployment insurance contributions for women over 56 and men older than 58 were waived (Bertelsmann-Stiftung Foundation, 2004b). By 2005, unemployed individuals would receive unemployment benefits for up to six months after which they could receive means-tested unemployment assistance. The *Kombi-Lohn* scheme was introduced on 1ˢᵗ February 2005 and targeted at those who faced particular disadvantage in the labour market such as individuals who had been unemployed for more than one year and were either under 25 or over 45 years of age. As low wages were considered to be a barrier to encouraging these individuals to move from unemployment benefits to paid employment, this scheme allowed individuals to receive half of their previous unemployment assistance as an in-work subsidy (with a threshold of €1,000 per month). This measure, it was argued, would reduce labour costs for employers, thereby allowing the creation of more jobs (Bertelsmann-Stiftung Foundation, 2005b).

Part-time work/pension arrangements

In 1993 a partial, transitional pension (*Gleitpension*) was introduced with the same conditions as the *Hacklerregelung für Langzeitversicherte* scheme and the individual received 70% of their pension amount when their working hours were reduced by 50% (OECD, 2006; MISSOC, 1996). New partial retirement options were made available in 1997 allowing individuals to work for 28 hours a week under the 'Solidarity bonus model' (*Solidaritätsprämienmodell*). The *Arbeitsmarktservice* (public employment service or AMS) also provided incentives for those who reduced their working hours so as to promote the recruitment of replacement labour (Federal Ministry of Labour, Health and Social Affairs et al., 1998). In 1999 gradual retirement schemes were improved to allow 'blocking' for employees who had worked 15 of the previous 25 years, had been with their current employers for a minimum of three months and were within five years of retirement. Under this scheme, an individual could reduce their working hours by between 40% and 60%. The employer then paid 50% of the difference between the wages which in turn was reimbursed by the state at 50% or 100% if the individual had been unemployed during the leisure phase. An individual was able to work full time for the first half of the agreed period, followed by zero hours for the second, thereby producing a lower *de facto* retirement age (OECD, 2005a).

The 1999 Pact for Older Persons (*Pakt für ältere ArbeitnehmerInnen*) also created two varieties of partial pensions or *Altersteilzeit*. The *Altersteilzeitgeld* meant that women over 50 and men over 55 could receive 75% of their previous earnings when they reduced their working hours by half. The employer received a subsidy equal to 25% of the gross earnings and was obliged to find a replacement worker to make up the missing hours. The *Beihilfe zur Förderung der Altersteilzeit* policy was similar but contained no replacement criteria, provided a smaller sum for the employee and required participants be two years older than the aforementioned *Altersteilzeitgeld* (EU-Employment Observatory, 2000). As part of this scheme, older workers were able to work between 40% to 60% of their normal working hours for up to six and a half years

(Federal Ministry for Labour et al., 1999). Individuals could then receive a partial retirement grant to promote gradual retirement (Committee of Regions, 2003). In addition, any time individuals over the age of 45 took out of employment would be counted as insurance months when calculating the final pension amount. As part of the *Pensionssicherungssreform* of 2003, the transitional pension (*Gleitpension*) was closed off as of 1st January 2004 due to the relatively low number of applicants. The *Beihilfe zur Förderung der Altersteilzeit* and *Altersteilzeitgeld* schemes were also closed to new entrants as of January 2010.

6.2 Belgium[1]

Though Belgium remained at the lower half of the table with regard to employment rates for older individuals, the rise from 2001 was 12.2%, well above the EU average, growing from 25.1 to 37.3% (Eurostat, 2011). Considering that in 2001 Belgium had the lowest employment rate for 55–64 year olds amongst EU15 nations, it had the greatest distance to travel towards the Stockholm target and therefore its progress in nine years has been comparatively substantial.

Pension policies

Pension principles and state pension age

The Belgian pension system operated on a PAYG principle for all employees. In the early 1990s, Belgium was the only EU15 nation aside from Sweden (and as of 1995 for certain cohorts, Italy) to have a flexible pension age, having introduced a system in 1991 to allow men and women to exit from the age of 60 and still receive a retirement pension (*pension de retraite/rustpensioen*) if they had made a minimum of 20 contribution years. In principle the retirement ages were set at 60 for women and 65 for men in order to allow the individual to have accrued sufficient contribution years for a full pension (40 years for women, 45 years for men). Therefore individuals who took the flexible option would experience a reduction of 1/40th and 1/45th of the pension amount for every year prior to state pension age (60 or 65) if they were female or male respectively (OECD, 2005b).

1997 saw a number of changes to the Belgian pension system, including the increase of pension ages for women in the private sector and public sector which were progressively raised from 60 to 65 by 2009, from 63 in 2003, to 64 in 2006, and 65 in 2009 (OECD, 2006; Whitehouse, 2007). In addition, the number of contribution years required for a full pension was also increased for women to become harmonised with men over this period (MISSOC, 1996, 1999–2005; OECD, 1997, 2005b).

In 2000, the career duration threshold that permitted exit with a full pension was at 42 years for women whilst for men it remained at 45. From this year onwards, the duration requirement for women was harmonised to that of men. The increase would be gradual: from 2000 to 2002, it would increase to 42 years; then to 43 years from 2003 to 2005; to 44 years from 2006 to 2009 and 45 years from 2009 onwards (MISSOC, 2000; OECD, 2005b). In addition, the pension age for women had been increased to 62 by the year 2000. The state pension age for women continued to increase to 63 by 2003. In addition, the contribution record for women who could receive the full pension was extended to 43 years (MISSOC, 1996, 1999–2005; OECD, 1997, 2005b).

Early retirement

As aforementioned, in the early 1990s, individuals were able to access the standard pension (*pension de retraite/rustpensioen*) at the age of 60 providing they were able to demonstrate that they had worked for a minimum of 20 years. However, the individual would incur a penalty of 2% per annum (OECD, 2006). From 1997, the contribution requirement for early access to the retirement pension (*pension de retraite/rustpensioen*) was increasing and by 2000, had risen to 28 years, and would continue to increase from 32 years in 2004 and 35 years from 2005 (MISSOC, 2005). In addition, as of this year, the penalty for withdrawing early was increased from 2% per year to 4.2% (OECD, 2006).

Deferred retirement

Belgium did not offer incentives for deferring state pension receipt, with the general principle that individuals should not work beyond the retirement age. Individuals could draw their pension at the same time as working, providing their earnings were below a certain threshold (OECD, 1997). Individuals who have not reached the requisite insurance years for a full pension can also continue working until they have made sufficient contributions (OECD, 2011).

Labour market policies

Early exit

In the early 1990s, Belgium had two main exit routes from the labour market that contributed to the low participation rate of older workers:

unemployment insurance for older unemployed individuals (*complément d'ancienneté*) and the conventional bridging pension for cases of dismissal (*prépension conventionnelle*) (Committee of the Regions, 2003).[2] As a general rule, in order to receive the latter pension, the individual had to be at least 58 years of age, have been made redundant and exited entirely from the labour market. The employer was obliged to replace the older person with an unemployed individual. If an industry was in economic difficulty, the employee could be 52 years old and the employer was not obliged to replace them. For enterprises declared to be in the process of reorganisation, the individual could be between 52 and 55 years of age and the employer was exempt from the replacement obligation (MISSOC, 1996). In these cases, a supplement (*indemnité complémentaire/aanvullende vergoeding*) of half of the difference between previous earnings and unemployment benefits was added to the conventional state unemployment benefits. This supplement was covered by the employer, whilst the state provided the unemployment benefit; thus the *prépension conventionnelle* was co-funded (MISSOC, 1996).[3]

In terms of the other main early exit route for older individuals who were unemployed with a contribution record of 20 years or more there was a supplement (*complément d'ancienneté*) after the first year on unemployment benefits. This ruling had been in place since 1991. In addition, since 1985, '*chômeurs ages*' i.e. individuals over the age of 55 (50 for those with reduced ability) who were unemployed for at least two years were exempt from job search. This age threshold was lowered to 50 and the unemployment requirement to one year in 1996 (OECD, 2006).

From the mid-1990s, the schemes that allowed employers to shed their older workers were reformed. In 1996, the aforementioned option of early exit via the conventional early retirement pension (*prépension conventionnelle*) in cases of dismissals was extended to those aged 50 in exceptional circumstances (OECD, 2005b). The early exit scheme that allowed employers to make those over 58 redundant if replaced by an unemployed individual was modified in 2000 to stipulate that the older person had to be eligible for unemployment benefits (*allocations de chômage*) (OECD, 2005b).

July of 2002 saw the repeal of the job search exemption for those aged between 50 and 57 who were newly unemployed; those already exempt would not have been required to re-register with the unemployment agency (Merla, 2004). Individuals with 38 years of contributions could still be exempt from the age of 50. The supplement for the older unemployed with a contribution record of 20 years (*complément d'ancienneté*) was still available.

Active labour market policies

In terms of active labour market policies for older individuals, since 1997 employer's social security contributions for older workers were reduced in certain sectors (the *Maribel* scheme). However, unemployment benefit supplements paid to the 50-plus were reduced for certain sub-groups below the age of 58 (Belgium, 1999). In 1998, there were policies in place to encourage the employment of older workers (OECD, 2005b). The progression-to-work programmes were available for those who had been unemployed for more than 24 months and provided them with the opportunity to work in the public sector and a recruitment advantage scheme whereby the Regions and the employer covered the wages of the individual (Belgium, 1999). From 1998, those over 25 and low skilled were allowed to access this programme earlier than the 24-month threshold (Belgium, 1999). In February 1998, Belgium introduced anti-ageism legislation to cover the field of employment.

In 1999 the *Maribel* scheme which lowered the social security contributions for those sectors subject to international competition was replaced by 'structural reductions in contributions' for those on low pay because the former scheme was deemed by the EU to interfere with fair competition (Vandenbroucke and vander Hallen, 2002). Simultaneously, in 1999 older workers were placed in vacancies that had proved difficult to fill (Committee of the Regions, 2003), perhaps to counter the creation of the category of 'older unemployed' (*chômeurs âgés*) who received additional unemployment benefits (OECD, 2005b, 2006).

As of July 2000, individuals over 45 and applying for work would warrant a reduction in their potential employers' social security contributions for five years under the *plan avantage à l'embache*. This scheme provided employers with a reduction to the social security contributions paid by employers for employees they recruited who had been unemployed for 12 months. If the individual had been unemployed for more than 24 months, the employer received a greater reduction (Belgium, 1999). In addition, for those older unemployed who re-entered the labour market at a lower wage than they were previously accustomed to, they would receive social security benefits in the event of unemployment based on these previous wages (Belgium, 2002).

Between 2001 and 2002, an inter-sectoral agreement was undertaken so as to improve incomes and working hours for those over the age of 45. The social partners also agreed to reduce the occupational stress of older workers and improve working conditions. Indeed, Vandenbroucke and vander Hallen (2002) argued that the social partners were following the example set by the state in terms of the removal of the policy which allowed older individuals to receive unemployment benefits without engaging in job search. This, they argue was key as "[i]t is the first time that so much attention has been paid in an inter-occupational agreement to a specific group of employees – real proof that the social partners are fully conscious of the problems resulting from excessively low activity levels in this age group" (Vandenbroucke and vander Hallen, 2002: 163).

In 2002, a scheme called '*Activa*' was introduced to replace the *plan avantage à l'embache*. As with the previous scheme, employers who recruited those over the age of 45 who had been unemployed for a long period of time received a partial exemption from social security contributions for a maximum of five years. These individuals were entitled to an employment subsidy as part of the programme (OECD, 2005b and Belgium, 2003). In addition, as a result of the age-discrimination legislation introduced in 2004, there was no longer mandatory retirement in the private sector (OECD, 2006). There has also been an increased focus on promoting the re-engagement of older workers into the labour market with targeted job search support and 'career check-ups' (Bredgaard and Tros, 2006).

Part-time work/pension arrangements

As of 1985, there was a partial career break scheme (*interruption de carièrre*) to allow those over the age of 50 to work part-time for three years. This was altered in 1993 to allow those aged 55 to work part-time until the age of 60 when they could have exited early. The employer was required to replace the individual with an unemployed person in order to receive a state subsidy (Lafoucrière, 2002: 51). The career break scheme was mainly utilised by those unable to qualify for other early retirement routes, principally women (Hemerijck et al., 2000).

There was also the option of a partial pension (*prépension conventionnelle à mi-temps*), introduced in 1994, for those over the age of 55 who would be allowed to work part-time whilst receiving a supplement covered partly by the unemployment insurance scheme and partly by the employer. The employer was also obliged to replace this individual with someone from the unemployment register. In 1999 part-time work was recommended as a means to gradually exit from the labour market and the narrow career break scheme (open to only 3% of the workforce) was replaced with a universal time-credit scheme, which could also smooth the retirement transition (Vandenbroucke and vander Hallen, 2002).

6.3   Luxembourg

Luxembourg, as with Belgium, still remained below the Stockholm target but had increased the employment rate for 55–64 year olds by 14%, rising from 25.6% in 2001 to 39.6% in 2010 (Eurostat, 2011).

Pension policies

Pension principles and state pension age

The state pension system in Luxembourg was insurance-based, requiring a minimum of 120 insurance months. The state pension age was set at 65 in Luxembourg, yet individuals could apply to the Ministry of Labour to continue working until the age of 68. In industries where there was a high level of youth unemployment, the Ministry could refuse an older individual's request (Grand Duchy of Luxembourg, 1998). In addition, the credit given for periods of sanctioned absences was generous with a maximum of nine years for higher education and ten years to cover the education of three children (OECD, 2005i).

Early retirement

Early retirement from the labour market in Luxembourg mainly occurred through two routes: first, as a result of the Law on Old Age, Invalidity and Survivor Pensions enacted in July 1987 and second, the Early Retirement Act of December 1990. The former allowed employees with over 480 contribution months to retire at either 57 years of age (or over 52 in the mining industry) under the premature old age pension (*Pension de vieillesse anticipée*) and the early access to the conventional pension (*Retraite anticipée*) respectively. The individual would receive no reduction to their pension amount. The 1990 Early Retirement Act focused on particular forms of work or occupations. Also, early exit could occur via a series of pathways. First, early retirement was granted for those on shift or night work at 57 years of age (*Préretraite des travailleurs postés et des travailleurs de nuit*). Second, early retirement for company restructuring (*Préretraite-ajustement*) allowed firms to retire workers at 57 years old or over and the *Prerétraite-Solidaririté* scheme permitted employers to do the same, provided the vacancy was filled by an unemployed individual. For those pensions which commenced at 57 years of age, at the age of 60 the individual would transfer to the *Indemnité de préretraite* which provided an amount that was 85% of the previous gross earnings for the first year, 80% in the second year and then 75% in the third year. Finally, employees could reduce their working time by between 40 and 60% as part of progressive early retirement (Committee of the Regions, 2003).

Deferred retirement

In 1995, there was the option of deferring pension receipt up until the age of 68 yet by 2005 this had been disbanded and instead the individuals' additional pension contributions would have been refunded at the end of the year. From this point onwards, individuals could work and receive their pension as it was mandatory to claim the pension at 65, but there were no additional incentives (OECD, 2007).

Labour market policies

Early exit

With regard to early exit, there was extended unemployment benefit durations for those over 50 years of age but these corresponded to contribution records (extensions of 12, nine or six months for 30, 25 or 20 contribution years respectively) throughout the entire time period under investigation.

Active labour market policies

Prior to the EU's 1997 European Employment Strategy, Luxembourg had implemented measures to aid older individuals' re-integration into the labour market under Law 23 07/1993, *Aide à l'embauche*. Under this scheme, the social security costs were reimbursed by the local authority over two, three or seven years depending on whether the individual was 30, 40 or 50 years old respectively (Committee of the Regions, 2003).

In Luxembourg, a roundtable on pensions (*Rentendësch*) was held in 2001 (OECD, 2005i). Subsequently, a scheme was introduced to facilitate older workers' reintegration into the labour market in 2002 (OECD, 2005i). Following a vote amongst the Unions in 2001, a law was passed in 2002 to incentivise longer working lives. Under this law, those over the age of 55 who had worked for 38 years could receive a double '*taux de majoration*' for the outstanding time in employment (Committee of the Regions, 2003).

Part-time work/pension arrangements

As aforementioned, the Early Retirement Act of December 1990 also introduced a scheme which allowed older workers to reduce their working time by between 40–60%. Following the 1998 NAP, this progressive early retirement system was simplified to allow those who were 57 and three years away from completing the requisite contribution record to exit at 60 to move into part-time work until this point.

In addition, if an unemployed individual was recruited to make up the working hours, the Employment Fund would contribute one-third of the wage cost. Similarly, a new scheme was added which allowed individuals over the age of 50 to work part-time. Again, the Employment Fund would pay the social contributions if the replacement came from the unemployment register. From 12th February 1999, a flexible framework was introduced to allow the social partners to organise their own working time (Grand Duchy of Luxembourg, 1999). In December 1999, Luxembourg introduced a measure to allow older workers to shift to part-time work in line with the EU's employment guidelines (Committee of the Regions, 2003). Under this scheme, an individual over the age of 49 could move to part-time work whilst the employment fund would cover the employer's social security contributions for this employee. In return, the employer would have to recruit an individual from the unemployment register to make up the hours. The social security contributions would be covered for seven years, unless the individual was hired on an indefinite contract whereby the state reimbursed the costs *ad infinitum* (OECD, 2001).

6.4   Summary and impact at the individual level

Both Belgium and Austria made significant moves to close off their early exit and retirement routes since the mid-1990s (and in particular Austria introduced stricter rules for younger cohorts); Luxembourg on the other hand retained theirs largely unaltered. The latter did however make moves towards encouraging longer working lives through the introduction of two partial pension schemes. All three opted to focus their ALMPs efforts on encouraging employers to retain and re-employ older individuals through incentives such as social security contribution exemptions and subsidies, as well as penalties for shedding this particular demographic from their workforce.

At the micro-level, in the case of Austria, desert was the key principle for decommodification and in addition, gender was salient with regard to age thresholds; thus if *Laurent* (55 with 35 years of employment contributions) had been a woman, decommodification options were available whilst irrespective of gender, *Jean's* (55 with a disjointed work history) contribution record prevented access to many routes. This presents a paradox as women are more likely to have incomplete work histories and the lower age threshold in effect give them less time to accrue sufficient contribution years. Nonetheless, this desert-focus meant that *Laurent* was able to access a great many early exit and early retirement routes whilst *Jean* would only have been able to retire early if disabled as a result of a work accident and would have been exempt from the job search requirement. However, if male, neither *Laurent* nor *Jean* could retire early unless they had been a miner but again this contained a contribution requirement. They would however have been exempt from the job search requirement of benefit receipt. Due to the unharmonised state pension ages, both *Sasha* and *Jude* would have been able to retire at 60 if female; if male, the data reflects Austria's desert-focus with *Sasha* able to retire early in a number of different scenarios (the Special Regulation for Very Long Insurance scheme (*vorzeitige Alterspension bei langer Versicherungsdauer*), the unemployment scheme with a 20 year contribution requirement (*vorzeitige Alterspension bei Arbeitslosigkeit*), the option for unemployed miners (*Sonderunterstützung Bergbau*) and those engaged in strenuous night work (*Sonderruhegeld*) and the option for those with a reduced working capacity (*Vorzeitige Alterspension aufgrund geminderter Arbeitsfähigkeit*)) whilst *Jude* was limited to the same options available to *Jean* (*Sasha* was 63 with 35 years of insurance contributions and *Jude* was also 63 with a disjointed work history).

Within five years, the decommodification options had begun to be curtailed for all model biographies. The scheme for those with a reduced working capacity (*Vorzeitige Alterspension aufgrund geminderter Arbeitsfähigkeit*) and the job search exemption were closed. As a result, all of *Jean* and *Jude's* options were removed and they were barred from accessing employment projects available to those who could have retired early. However, they could have benefited from other elements of the range of ALMPs introduced, particularly those that afforded older individuals protection from redundancies and employers incentives for their recruitment. *Laurent* and *Sasha's* early retirement options were also retrenched through the addition of penalties to all the early retirement schemes, with the exception of the options for miners and those in arduous labour. Labour market participation was also encouraged for females in this age cohort through the introduction of incentives for the deferral of state pension. The partial pensions were made more accessible with the reduction of the contribution requirement.

Decommodifying policies were retrenched further by 2010 with the increase of age thresholds, thereby limiting *Laurent* to the same routes as *Jean* and *Jude* (for certain occupations); if female, *Laurent* could no longer access the Special Regulation for Very Long Insurance scheme (*vorzeitige Alterspension bei langer Versicherungsdauer*) as the age threshold had been raised to 56.5 for women and 61.5 for men. Indeed, *Sasha* would only have been able to access one additional route for long insurance durations though this option was being closed off gradually by 2017 and the option for long-term unemployed individuals (*vorzeitige Alterspension bei Arbeitslosigkeit*) had also been closed off. They would however have still been able to access the option for unemployed miners (*Sonderunterstützung Bergbau*) and those engaged in strenuous night work (*Sonderruhegeld*). Special policies had been introduced for individuals over 60 to assist their labour market participation and the incentive for deferral was increased. In addition, by 2010, a different system had been introduced for those who were under 50 in this year. This system was flexible, thereby moving away entirely from the desert-focused system to a rights-based model of decommodification. Individuals would therefore have autonomy over their exit, though their choices may be constrained by their contribution record as this was directly linked to their pension amount. The partial pension was by this year closed off.

Belgium had in place a mix of labour-market and desert-focused schemes in 1995 with regard to the model biographies. Belgium began the ten year period with differential state pension and early retirement age thresholds for men and women, the latter of which was gradually raised. In addition, in 1995 individuals had the option universally of exiting at 60 years of age if they had made 20 years of contributions with a reduced pension amount unless they had made 40 years of insurance contributions. With regard to the remaining routes, early exit in the form of an extended benefit duration would only have been accessible to *Laurent* and *Sasha* if male (if female they would have retired at 60) due to their contribution record whilst the remaining options were available to all model biographies in instances of economic difficulty or firm restructuring.

By 2010, the state pension age for women had been raised to 64 and early exit in the form the job search exemption was retrenched as the age threshold was raised to 57 and closed to new applicants. As a result, *Jean* and *Jude* could only exit via the two schemes for companies in economic difficulty; the requirement for the flexible scheme of 20 years had been raised to 35 years and the reduction per year prior to the state pension age had been increased to 4.2%. Thus in the case of Belgium, gender and contribution records became less significant in the bestowal of penalty-free decommodification whilst the needs of the market retained salience through the lack of retrenchment of the early retirement schemes for industries wishing to restructure.

Luxembourg retained the same policy approach to *Laurent* and *Jean* throughout the entire period due to the fact that the majority of their decommodification policies contained age thresholds higher than 55. However, extended unemployment benefit durations in Luxembourg were contingent upon contribution records as opposed to age, meaning *Laurent* would have had been able to have been decommodified for longer than *Jean*. In terms of early retirement options, age was key to preventing both *Laurent* and *Jean* from accessing the routes available to both *Sasha* and *Jude*. These options were labour-market-focused in that they applied in instances of redundancy to allow firms to restructure and for certain kinds of work. The desert-focused options had contribution thresholds that were too high for both *Sasha* and *Jude* to access (this and the age threshold meant *Laurent* was barred).

These three nations all had quite distinctive decommodification approaches. In Austria, the focus in 1995 was on contribution records. Gender was key in that the age thresholds for men and women were unharmonised, both giving the latter the opportunity to exit early but also less time to accrue sufficient credits to retire early. There were a comparatively large number of early retirement options available in 1995 but by 2010, these had been retrenched through the increase of age thresholds and reductions to pension amounts for those leaving the labour market prior to the state pension age. Active ageing was therefore being encouraged, even more so for those under 50 in 2005 for whom a new flexible system was in place, encouraging longer labour market participation. Belgium differed insofar as though they too had desert-focused schemes which were also retrenched, but there were also labour market focused options for those employed in industries in economic difficulty which were not reformed by 2010. Thus labour market participation was still key to penalty-free decommodification, but less so in terms of long, uninterrupted careers while employment in particular industries could still allow for early exit and retirement. Luxembourg too focused on desert and labour market conditions for decommodification policies but did not retrench these over the entire period. These options did however have a higher contribution requirement than in other EU15 nations and there were no early retirement options for those under 57 years of age, only extended benefit durations but these were contingent on contribution records.

7
Group IV: The Laggards – Slow Progress Towards Stockholm

As aforementioned, seven EU15 nations remained below the Stockholm target in 2010, yet some had made significant progress when their low employment rates for the 55–64 age group at the beginning of the decade are taken into account. However, Greece, Spain Italy and France still remained below the target (significantly so in the case of France) and the progress they had made was less than the average for all EU15 nations (9.6%).

7.1   Greece

Greece, as with Belgium, Luxembourg and Austria, remained both below the Stockholm target and the EU15 average for 55–64 year olds employment (48.4%). However, unlike the aforementioned three nations, the employment rate for older people in Greece too did not increase at the same rate, only rising by 4.1% from 38.2 in 2001 to 42.3 in 2010 (Eurostat, 2011). However, Greece's dual early retirement system for different birth cohorts will potentially increase this rate in the future (see *Early retirement*).

Pension policies
Pension principles and state pension age

Greece's pension system operated on a PAYG principle with benefits based on employees' and employers' contributions. Due to the coverage of this system, private pension schemes were relatively underdeveloped. Though the state controlled the pension system, there was a great deal of fragmentation in terms of the sub-schemes available for the various sectors broadly spilt into Primary, Auxiliary and Provident funds. Law 1902/90 in 1990 had set the retirement ages at 60 for women and 65 for men and required a minimum contribution record of 4,500 insurance days for a full pension (Triantafillou, 2007).

Major reform of the pension system took place in 1992 to tighten pension eligibility, effectively creating two systems: one for those insured before 31st December 1992 and one for those who entered the labour market after this point. Law 2084/1992 modified pension calculations but still allowed early exit prior to 65 years of age (Committee of the Regions, 2003). The gendered state pension ages were harmonised in 1993 to 65 years of age for those entering the labour market in this year and the early retirement schemes were altered, as will be outlined below (OECD, 1997).

Early retirement

In terms of the early pensions, prior to the reforms of 1992, men could exit at 60 years of age if male and 55 if female and employed in arduous or unhealthy work with a full pension, or universally with a reduction of 6% per year prior to the state pension age. For those in the construction industry, the age thresholds were 57 for men and 52 for women (Triantafillou, 2007). As part of the 1992 reforms, a dual pension system was created with new rules for those entering the insurance scheme after 1992. For those entering before this date, changes were also made to the early retirement schemes. New routes were added, including a scheme whereby individuals could exit from 62 or 57 if male or female respectively if they had 10,000 days insurance, which could then be lowered to 58 with 10,500 days.

For those insured after 1992, the age thresholds were harmonised for men and women at 60 in cases of arduous labour as well as if the individual had worked for 20 years and had children who were underage. These individuals would not face a reduction to their pension amount. However, there were also schemes with reductions (6% per year) including one for women aged 50 with a disabled/minor child and 20 years of insurance and another which allowed women with three or more children and 20 years of insurance to bring forward their retirement by three years per child (with a limit of 50 years of age) (MISSOC, 1994).

By 2000, the early retirement schemes were again modified. With regard to those employed before 31st December 1992, the scheme for long insurance durations was unreformed yet the policy for arduous labour had a contribution requirement added, specifying 4,500 working days of which 3,600 were in arduous labour and 1,000 were worked within the previous ten years. The universal scheme which allowed people to retire at 60 if male and 55 if female with a reduction of 6% per year had a contribution requirement of 4,500 days of which 100 were in the previous five years added. A new scheme was created for women with dependent children and 5,500 contribution days who could exit at 55 years of age, or at 50 with a reduction of 6% per annum (MISSOC, 2000). In addition, those who had completed the requisite 10,000 days of insurance (of which 100 per year during the last five years) or 10,500 working days (of which 7,500 days were in arduous or unhealthy conditions) but wished to retire even earlier than the aforementioned thresholds could retire at 60 years for men and 55 years for women (instead of 62 and 57) or 53 years for both men and women respectively but they would have incurred a 6% per annum reduction to their pension amount.

The options for individuals employed after 31st December 1992 had been curtailed by this year. In terms of an unreduced pension amount, this would only have been possible at 60 for those with 4,500 insurance days of which 3/4 were spent in arduous labour or for mothers of disabled or minor children who could also have exited at 55 with 6,000 insurance days. In terms of a reduced pension, those with 4,500 insurance days with 750 in the previous five years could exit at 60 years of age with a loss of 6% per year (MISSOC, 2000).

Five years later, the early pensions had been altered. For both cohorts there was the option of exit at any age with 37 years of contributions. In addition, an incentive for deferral was introduced in terms of a 3% annual increase per year worked post-state pension age. The remaining early retirement options remained unchanged until Law no. 3655/2008 raised a number of the age thresholds for the early pensions available to those insured before 1992. For the scheme available to those with extensive contributions aged 58, the age threshold is gradually being raised from 58 to 60 from 2013 to 2016. Over this period, the early retirement option for those in arduous labour is also being raised from 55 to 57. The scheme for women who had made 10,000 days' worth of contributions was also being raised from 57 to 60 from 2013 and 2018. The age limit for mothers was also being increased from 50 to 55 between 2013 and 2017 (Greek Ministry of Labour and Social Security, 2008).

Deferred retirement

There were no incentives for deferral in place pre-2005 (Triantafillou, 2007). As of 2005, an incentive for deferral was introduced in terms of a 3% annual increase per year worked post-state pension age beyond 35 years of contributions until the age of 67 (MISSOC, 2006). From 2007, this increase had been raised to 3.3% and after 2008, the age threshold was increased to 68 (MISSOC, 2007, 2008; OECD, 2011).

Labour market policies

Early exit

Throughout the period of 1995 to 2010, the Greek unemployment benefit system remained unreformed. Individuals over the age of 49 with a certain level of contributions could receive unemployment benefits for the extended period of 12 months and 210 days. All individuals received an additional three months of benefit at a reduced rate, having exhausted the full rate.

Active labour market policies

With regard to ALMPs, in 1995 job subsidies were created for those over 54 years of age (Mestheneos, 1998). In Greece, measures to encourage the employment and retention of older workers were introduced from the mid-1990s onwards. In 1997, a range of active ageing policies were set up, including the creation of 50,000 full-time jobs for all age groups and employment cards which entitled an unemployed person to training, employment subsidies or self-employment. There were also one-year subsidies for employers recruiting unemployed people on the condition they kept that person on for a minimum of between three to six months (Mestheneos, 1998). January 1997 saw the launch of a law to remove the age limit of 46 to enter the Manpower Employment Organisation's (*Οργανισμός Απασχολήσεως Εργατικού Δυναμικού*, OAED) rapid training courses (Mestheneos, 1998). A month later, a programme called 'Cell of work re-integration' was initiated to focus on those workers who had been made redundant in the event of a company restructuring. It provided opportunities for education, training or self-employment. For those men over 55 and women over 50, access to special funds could be received for three years (Greek Ministry of Labour and Social Security, 1998). New measures were also introduced for this group as part of the Programme for the Exclusion from the Labour Market. The general aim of these measures is to ensure those over 45 "are competitive – from a skills viewpoint – in relation to young people, they can adapt rapidly to the needs of the production process and also in order to avoid the firing of older employees because of lack of skills" (Greek Ministry of Labour and Social Security, 1998: 49).

A scheme was launched in 1998 called the Integrated Intervention Programme which strove to ameliorate large-scale redundancies as a response to economic difficulties. The aim was to prevent those out of work from drifting into long-term unemployment by providing active support in the form of a skills assessment of the cohort made redundant and attempts to place this group back into the local economy. Failing this, support would be provided in the event of relocation. These individuals would also receive training to the end of re-entering employment or becoming self-employed. The unemployed individual would be able to make use of this support for two years (Greek Ministry of Labour and Social Security, 1999).

Following a pilot in 1998, the OAED introduced a programme focused on those unemployed individuals over the age of 58 in 1999. The OAED launched a programme to be financed by the Employment and Vocational Training Fund (*Λογαριασμό για την Απασχόληση και την Επαγγελματική Κατάρτιση*, LAEK) in 1999 to subsidise employment for those who were unemployed and approaching retirement age. Social partners also made contributions into the fund. If employment could not be found for an individual in this situation over 60 if male and over 55 if female, they would have received insurance until the state pension age. This programme ran until 2002 (Greek Ministry of Labour and Social Security, 2000).

The 1999 introduction of the Operational Programme 'Combating exclusion from the labour market' was remodelled to include needs-based training and measures which focus on employment promotion (Greek Ministry of Labour and Social Security, 2000). The Elective Studies Programmes provided older individuals with lifelong learning oppor-tunities, including the chance to engage in tertiary education or retrain at any point in their working lives (Greek Ministry of Labour and Social Security, 2000).

In 2001 a new measure was introduced to focus on the long-term unemployed aged 45 to 65 to promote labour market re-integration. These individuals, who had been unemployed for an excess of 12 months, had exhausted their regular OAED benefits and had a family income below a certain threshold, could receive a subsidy of around 30% of an unskilled worker's wage for up to 12 months. The subsidy was sus-pended in the event of training or education. Several of the measures already instigated were continued in 2001, including those that pro-vided vocational or in-house training and the LAEK Fund programme to subsidise the employment of unemployed individuals approaching retirement. As of 2001, these measures were expanded to provide voca-tional training in new technologies, not only for the unemployed, but in addition those in work. Also, employers who recruited those over the age of 55 who had a minimum of 6,000 insurance stamps would only pay 50% of the insurance contribution for these individuals. The state paid the difference through the IKA fund (Greek Ministry of Labour and Social Security, 2002). In 2003, the OAED launched new subsidies for older individuals. A subsidy of €17.61 per day for the first two years and €20.54 for a further three years was provided for those five years under the state pension age who still had to fulfil 1,500 social security credits. Additionally, businesses providing employment for those over the age of 45 would be provided €21.5 per day for two years (Kathimerini, 2003).

In terms of anti-ageism legislation, regulation 43/78/2000 included age with alongside racial or ethnic origin, religious or other beliefs, disability, age or sexual orientation. The Law number 3304/2005 implemented the Directives 2000/43/EC and 2000/78/EC, stating in Article 13 paragraph 3 on the defence of rights points out that "legal entities which have a legitimate interest in ensuring that the principle of equal treatment is applied regardless of racial or ethnic origin, religious or other beliefs, disability, age or sexual orientation can represent the person wronged before any court and any administrative authority with the written consent of the person wronged" (in Ktistakis, 2005: 3).

A number of measures were introduced as part of Law 3369/2005, including the expansion of Second Chance Schools and Adult Education Centers, the extension of Life-Long Learning Institutes in Universities and the formal recognition of skills and experience (Greek Ministry of Labour and Social Security, 2008). From July 2007, subsidised employment was launched for 20,000 people, with an emphasis on women over the age of 45 and men older than 50 (Greek Ministry of Labour and Social Security, 2008).

Part-time work/pension arrangements

There is no option for a partial pension receipt in Greece.

7.2   Spain

Spain, as with Greece, did not exceed the Stockholm target or the EU15 average employment rate for those waged 55–64, and only made a small gain of 4.4% from 2001 (39.2%) to 2010 (43.6%) (Eurostat, 2011).

Pension policies

Pension principles and state pension age

Spain too adopted an insurance-based pension system for employees, in addition to a flat-rate, means-tested pension for those who had failed to make sufficient contributions for the former scheme (Chuliá,

2007; Whitehouse, 2007). The legal age of retirement in Spain prior to 2002 set at 65 with a minimum seniority requirement of 15 years, and 35 for a full pension (Araico, 2004). The changes with regard to the flexibilisation of the state pension age are discussed below under *Early retirement*.

Early retirement

There were three main early retirement routes: the first applied to those with 15 years of social security contributions, some of which had to have been made prior to 1967, who could exit at 60 with a reduction of 8% per year prior to retirement (Araico, 2004). The second and third routes contained no reduction and were for times of industrial restructuring and individuals in arduous employment. With regard to instances of industrial restructuring, there were two sub-types of pre-retirement schemes. First, workers aged between 55 and 60 could receive contributory unemployment benefits for two years and welfare benefits for three years from the National Institute of Employment (*Instituto Nacional de Empleo*). The company in this case would cover up to 75% of the previous wage and the state provided a subsidy in accordance with the amount of reduced pension. Second, if a company was in a fiscal crisis, it could make individuals over the age of 60 redundant. These individuals could receive 75% of the basic income upon which social security contributions were calculated. This salary was covered 60% by the employer and 40% by the state. These individuals would receive an income made up of compensation from their employer, unemployment benefits and a state subsidy. Individuals could also exit with no state subsidies (aside from unemployment benefits and pensions) if based on collective bargaining with a large employer (OECD, 2005l).

For individuals employed in arduous labour (such as bullfighters, transportation workers and the like) early retirement was also permitted (Boldrin et al., 1999). In the case of the coal mining industry, those over 52 years of age were entitled to contributory unemployment benefits for two years, after which they received welfare benefits. As a supplement, the Ministry topped up the workers' income to 78% of the previous gross wage (OECD, 2005l).

From 1985, an individual aged 64 planning retirement could work alongside their replacement from the unemployment register part-time under the '*Relevo*' (take-over) scheme (Forteza and García-Zarco, 1998). The substitution contract also allowed firms to retire an individual early providing their replacement was recruited from the unemployment register. The employer received a reduction of 50% of the social security contributions for the new employee (Araico, 2004).

1997 was a key year of reform in Spain, following 1995's *Toledo Pact* which pledged to reform the social security system to become more equitable and balanced. In 1997, the penalty for early exit was decreased in 1997 to 7% for those for whom exit was not voluntary, whilst for those who left of their own accord it remained at 8% (MISSOC, 2001). In 1999, the age threshold was lowered to 60 years of age for the *Relevo* scheme (Araico, 2004).

In 2002 as part of the Law 35/2002 reforms, retirement was made more flexible in Spain (OECD, 2005l). Prior to this year, individuals who had contributed for 15 years to the system pre-1967 could have exited before the state pension age. This pre-retirement measure was expanded to include those with a minimum of 30 years of contributions, not necessarily pre-1967, who could exit between the ages of 61 and 65 if unemployed for six months for reasons 'outside of their control' (European Commission: Economic and Financial Affairs, 2005a; Spain, 2001, 2004; Committee of the Regions, 2003; OECD, 2005l). However, those individuals' pensions were reduced by between 6–8% per year before the retirement threshold. The reduction corresponded to the individuals' contribution years with those with 30 years of contributions incurring a loss of 8% per year; those with between 31 and 34 years receiving a deduction of 7.5%; 7% for those with 35 to 37 years; 6.5% for those with 38 to 39 years and for those who had made 40 and above contribution years, the reduction was 6% per year short of the state pension age (European Commission: Economic and Financial Affairs, 2005b). Therefore, those who exited at 61 received a pension reduced by between 24–32%. In addition, for those who deferred pension receipt beyond 65 years of age and had worked for 35 years, there would have been an increase of 2%. Thus early exit without a reduction was no longer possible, unless the individual worked within central government or was covered by laws regarding industrial restructuring. As part of the flexibilisation of pensions, an individual was able to combine part-time work and a part-time pension. These individuals would receive the pension amount in proportion to the reduction in working hours (OECD, 2005l).

Deferred retirement

Individuals working beyond the state pension age of 65 would increase their pension by 2% per annum (Araico, 2004). As of 1 January 2008 following Law 40/2007, of 4th December, if the individual had made 40 years of contributions, each additional year would incur an increase of 3% (Social Security Administration, 2008; Spain, 2008).

Labour market policies

Early exit

With regard to unemployment benefits, there were two systems. The first, the contributory system was for those made redundant who were actively seeking work. The limit for receipt was between four and 24 months, depending upon the individuals' contribution record. They would then receive 70% of their previous earnings for six months, reduced to 60% of the reference earnings for the remaining period. When the entitlement to this scheme was exhausted, the individual could access the assistance benefit scheme for between three to 30 months. In addition, for those who had exhausted their rights to unemployment insurance, there was special provision for individuals over the age of 52 if they had contributed to the unemployment system for a minimum of six years and the pensions system for the required minimum. These social plans provided a *de facto* bridge to pension access. Individuals could receive this benefit for 14 years and was set at 75% of the national minimum wage (OECD, 2005l). In addition, the Industry Law scheme (*Ley de Industria*) provided individuals aged 52 to 60 with contributory unemployment benefits for two years, followed by welfare benefits plus a supplement from their employer.

Active labour market policies

In Spain, older workers had specific protection under labour law (Drury, 1993). The Spanish labour law of 1980 made specific reference to older workers thus: "workers have the right not to be discriminated against in their application for employment or once employed on grounds of sex, marital status or age, within the limits set out in this Law" (in Drury, 1993: 56). Indeed Article 35 of the Spanish constitution (1978) states "none shall suffer any discrimination whatsoever on grounds of birth... or any other personal or social condition or circumstance" (in Drury, 1993: 56) and also incorporates the individuals' right to work (Forteza and García-Zarco, 1998). Older individuals were however, exempted from the job search requirement of unemployment benefit receipt.

As part of the reforms of 1997, to remove disincentives for recruiting older workers, the severance pay for those over 45 was reduced, as were employers' social security contributions for older individuals they recruited (OECD, 2005l). The reductions were awarded on a sliding scale with those aged 45 to 54 receiving a reduction of 50% for the first year and 45% thereafter if male and 60% and 55% respectively if female. For employees aged between 55 and 64, the reduction was 55% and thereafter 50% for men and 65% and 60% for women. In addition, firms with workers aged over 60 years of age with a minimum of five years of seniority also received a reduction to their social security contributions (OECD, 2006). Though not directly excluded, older workers were not the focus of governmental training programmes which instead catered for the young unemployed. In Spain, new measures were introduced that specifically targeted older women and the self-employed in this year (Taylor, 1998), offering vocational skills training, career guidance and social support.

From 1998 to 2000, the Continuous Professional Training Programme ran for employed individuals, particularly focusing on certain groups including those over 45 years of age (Spain, 1999). Employment Workshops ran for those over the age of 25 who had been unemployed and had difficulties finding work. These workshops lasted between six and 12 months. From 1999, these workshops were specifically targeted at those over 40 years of age (Spain, 1999). Individuals were also able to use replacement contracts of between 30% and 77% of the working day (Spain, 1999). The Programme for Promoting Work was established in 2000 to provide allowances for self-employed individuals who employed an individual over the age of 45 under a permanent contract as their first employee (Spain, 2000).

To promote the employment of certain groups over-represented amongst the ranks of the unemployed, Spain introduced a reduction for employers recruiting certain 'disadvantaged' individuals. Those over the age of 45 and unemployed were one of the groups targeted by this measure from March 2001 onwards. In addition, the compensation pay which was deemed a disincentive for employers was modified. For those on temporary contracts, the compensation was reduced to eight days' pay per year employed. In addition, the indefinite contracts introduced in 1997 which decreed the cost of unfair dismissal should be limited to 33 days salary for each year worked with a ceiling of two years' pay were extended to cover the entire working population with the exception of those aged between 30 and 45 (Bertelsmann-Stiftung Foundation, 2001a). Social plans aimed to minimise the disproportionate redundancy of older workers by promoting part-time contracts as an alternative in 2001 (Spain, 2002). The Active Job Finding Programme was extended in 2001 to include those over the age of 45. Individuals enrolled on this programme could receive unemployment benefits while engaged in part-time voluntary social work or work for non-profit organisations (Spain, 2001).

Until 2002, older individuals were exempted from actively seeking work. In May 2002 the unemployment protection system was reformed in Spain to promote 'activation' (Bertelsmann-Stiftung Foundation, 2002c). From this point onwards, individuals had to make a written commitment to accept help with job search and suitable employment (*empleo adecuado*) in exchange for unemployment benefits. 'Suitable employment' referred to a job that the individual had performed at another point in their working life for at least six months. Those who had not found a job after 12 months of benefit receipt were obliged to accept a job offer following training and if this job was low-paid, they received a reduction to their social security contributions. Further conditions applied to the geographical mobility of the unemployed individual, obliging individuals to take any job within 30 kilometres of their home so long as the cost of the journey did not exceed 20% of their wages. For those individuals who refused a job offer, penalties ranged from the loss of between one month's benefit to the total withdrawal. Additionally, assets and investments will negate an individuals' right to unemployment benefits. This is particularly pertinent for those over the age of 52 and receiving early retirement benefits in addition to those receiving redundancy pay. Following a general strike, this was modified to allow those receiving redundancy pay to also claim benefits. Under these reforms, the 'integration contract scheme' was expanded to include all individuals over the age of 45 who have been unemployed for one month (OECD, 2005l).

Following the agreement with the two main unions, a series of measures were outlined in 2001 and subsequently introduced in 2002 as part of Law 35/2002. First, mandatory retirement was prohibited in the public sector. In addition, to encourage the recruitment of older individuals, employers' social security contributions for workers over 60 on a permanent contract with five years of seniority were reduced by a minimum of 50%. This reduction would increase by 10% each year and could reach 100% by the time the individual reached 65. In addition, for workers with over 35 years of contributions and over the age of 65, employers were entirely exempted from contributions. To discourage the dismissal of workers over the age of 55 in times of economic restructuring (*Expedientes de Regulación de Empleo*), employers were obliged to cover the pension contribution of these workers until they reached the age of 61 (OECD, 2005l).

On 30th September 2003, the *Toledo Pact* reduced or removed the social security contributions for those 'high risk groups in the labour market' (Spain, 2004). Also in this year age discrimination legislation was launched with regard to employment, despite the reservations of the Spanish Confederation of Employers' Organisations (CEOE) (Taylor, 2005).

In 2006, the agreement for the Improvement of Growth and Employment was signed by the government and social partners and led to Law 43/2006 which sought to address the goals of the Lisbon agenda and therefore included measures that focused on older unemployed individuals. Employers recruiting people over the age of 45 could receive a reduction to their social security contribution of €1,200 per year for the duration of the contract. In order to also encourage the retention of older workers, employers would receive a 45% reduction to their social security contribution for employers over the age of 59 who had been with their company for at least four years; this increased by 10% each year until it reached 100% at the age of 65 (Spain, 2008).

Part-time work/pension arrangements

In Spain, a partial retirement programme, *pensión de jubilación parcial*, was available since the 1960s to those over the age of 62 and these individuals were able to retire fully at 65 with the additional credits accrued from this period in part-time work (Forteza and García-Zarco, 1998). The scheme initially aimed to promote job creation and an additional criterion of a 'substitution contract' whereby an unemployed individual had to be recruited to make up the hours of the partial retiree was introduced in 1984. In 1999, the rule which stated an individual had to work 50% of their original contracted hours was changed to between 30–67%. The individual had to have made 15 years of contributions, of which two had been made in the last 15 years. In 2002, further reforms were undertaken to state that the working time could be reduced to between 25% and 85% of full working hours. The remaining hours still had to be filled by an individual from the unemployment register, unless the retiree was over the age of 65 as this reform expanded the partial retirement option to this age group. In the hope of incentivising the extension of working lives, the individual was still able to accrue pension credits through this part-time work post-state pension age, and employers were exempted from their social security contributions (MISSOC, 1999; Social Security Administration, 2006; Belloni et al., 2006).

In 2007, Law 40/2007 reformed the partial retirement scheme, as the Spanish government felt the previous rules had been exploited. The age threshold was lowered to 61 and the scheme was limited to those who had entered the social security system after 1st January 1967, to be effective as of 2009; those who entered the system prior to 1967 could still exit before this date. Individuals had to have made 30 years of contributions and have been with their employer for a minimum of six years to be eligible. The working hours could be reduced between 25% and 85% (Spain, 2008).

7.3   France

France too had not reached the Stockholm target by 2010, and had only made modest progress towards it since 2001 (+7.8%). France's employment rate for older people grew beyond 30% in 2001, but had not reached 40% by 2010 (Eurostat, 2011). As a result, Austria which had had lower employment rates in 2001, had now overtaken France, and other nations such as Luxembourg and Belgium were moving closer.

Pension policies

Pension principles and state pension age

The French pension system was extremely fragmented and occupation-ally segmented. In general, the French pension system operated on a PAYG principle, financed by employers and employees, and was divided into three schemes for private, public and self-employed individuals. The basic state pension, the General Scheme for Employees (*Régime Général d'Assurance Vieillesse des Travailleurs Salariés*, RGAVTS) provided the full pension amount when 160 insurance quarters had been reached, at the age of 65 or in particular instances (for example if the individual was unable to work). In the private sector, an individual's pension amounted to 50% of their average wages taken from the best ten years of their career in instances where they had made the requisite 37.5 years of contri-butions. If the individual exited early, their pension would not only be reduced due to the prorating effect, they would also incur a loss of 10% per missing year with a maximum of five years (as introduced in 1983). The rules differed for the public sector with individuals receiving 75% of their monthly wage for the previous half year, with early exit incurring a loss in line with prorating only (OECD, 2005e).

   In addition, there were two compulsory occupational schemes: the General Association of Retirement Institutions for Executives (*Association Générale des Institutions de Retraite des Cadres*, AGIRC) established in 1947; and in 1961, all workers in the private sector were required to join the pri-vate Association of Complementary Retirement Systems (*Association pour le Régime de Retraite Complémentaire des salaries*, ARRCO). The retirement age for these schemes was 65, with penalties applied to those retiring from the age of 60 onwards (Deville, 2000). These two schemes provide an additional pension amount which makes up a small proportion of the total income post-retirement (Milner, 2007).

Ebbinghaus (2006: 159) classes the AGIRC and ARRCO as "collective schemes negotiated by the social partners provide (quasi) second-tier pension in France", as is the case of the Netherlands. Though a 'quasi' second-tier, Ebbinghaus nonetheless includes both in the section on private occupational pensions that were provided by employers and social partners in his book. Barbier and Théret (2000: 30) concur that these schemes are part of a 'quasi-integrated system' due to their compulsory nature. There is no state-provided second tier; instead there are the mandatory occupational schemes. In addition, the state's role in the two schemes is considered low and the management, rules and implementation are undertaken by the social partners (IPOS, 2006; Barbier and Théret, 2000). However, any changes to these schemes have to be approved by the Minister of Labour. Yet for the International Organisation of Pension Supervisors (IPOS), these schemes are part of the private pillar in spite of their compulsory nature.

However, other authors see the French pension system as divided into three pillars: compulsory state pensions, occupational pensions and private pensions. The compulsory state pensions are in turn divided into an earnings-related state pension and occupational schemes, run by the ARRCO and AGRIC. Kalisch et al. (1998) argue if pension schemes are categorised according to their level of programme coordination, the ARRCO and AGRIC schemes should be included in the public first pillar. Indeed "as the basic element, public pension programs are identified as those which are managed by public entities and/or with a great extent of national-level co-ordination... those schemes with national-level financial co-ordination based on pay-as-you-go funding... are regarded as public schemes in the System of National Accounts even if they are primarily managed by the private entities" (Kalisch et al., 1998: 65). In addition, "although schemes are not run or financed directly by the state they are regarded as public pensions" (PPI, 2003: 21). However, in terms of this book, it is precisely because these schemes are neither state-financed nor managed that they are not considered state-provided decommodification policies. France also presents an interesting case as there was also a comparatively significant amount of early exit and retirement schemes which were also established and financed by social partners as opposed to the state. In comparison to the Netherlands with the VUT scheme, France had a great many more of these non-state provided early exit and

retirement options which were, as a result, insulated from retrenchment. As a result, Guillemard and Argoud (2004) argue retrenchment
to early exit and retirement has been offset by actions of social partners
as part of National Unemployment Insurance Administration (UNEDIC).

In 1983, the retirement age for the General Scheme for Employees
was lowered to 60 years and the requirement for a full pension was
raised to 37.5 years of contributions (Argoud and Guillemard, 1999).
The 1993 Balladur reform gradually increased this to 40 years by 2003
for private sector employees; the subsequent Fillon reform of 2003
increased this to 40 years for public sector workers by 2008. By 2012, it
will again increase to 41 years for both sectors, and will further rise in
line with rises in life expectancy (OECD, 2007; Belloni et al., 2006).
Significant reform of the state pension age was undertaken in 2010 to
increase it by two years initially over a period of six years from 2012
(rising by four months per year to reach 62 in 2018).

Early retirement

Early retirement schemes in France were introduced in the 1960s as a
tool to adjust to economic changes by allowing for industrial restructuring. As aforementioned, France is distinct in terms of the number of
early exit and retirement options available through the social partners
who were in agreement that early exit would provide opportunities for
young, unemployed individuals (Gendron, 2011). The National Employment Fund (*Fonds National pour l'Emploi*, FNE) was created in 1963 and
administered a scheme to allow for restructuring, with firms who had
signed an agreement with the state permitted to retire individuals at the
age of 60 (Guillemard, 1991). Older workers however, could not receive
redundancy pay or become re-employed. Within ten years, early retirement was viewed as the right of the individual worker, providing autonomy over their labour market exit (OECD, 2005e; Gendron, 2011) and
employers, trade unions, government and older workers themselves had
been in agreement that early exit was essentially a good thing for all concerned. As a result, early exit was the predominant model in France; the
traditional route from work to pension was only used by a minority in
France (Guillemard, 1993: 38).

In the 1980s, a number of new schemes were created including the
modification of the aforementioned FNE-administered scheme. The
new Special National Employment Fund Allowance (*Allocation spéciale
du Fonds national de l'emploi*, ASFNE) came into effect in 1981 and allowed
individuals dismissed from firms that had signed an agreement with
the state to exit aged 56 years and two months old (in exceptional

cases, this could be lowered to 55) with 70% of the gross wages (Guille-mard, 1991). The individual had to have made ten years of insurance con-tributions and been with the firm for a year (Jolivet, 2002; Guillemard, 1991; Eichhorst and Rhein, 2005).

In 1993, more stringent conditions were applied to the ASFNE, with the payment reduced to 65% of their gross salary (subject to a ceiling) and access became increasingly restricted, limited to those who had been made redundant due to economic difficulties and those over the age of 57 for whom re-employment was unlikely (France, 2003). The Balladur pension reform of 1993 also introduced a *décote* (dis-count) of 10% per year for those exiting prior to the state pension age in the private sector (Belloni et al., 2006).

Due to demographic changes, early exit in France has become prob-lematised and prompted the government to act. Their actions, how-ever, were somewhat undermined by the social partners who began to create their own exit options; indeed, the French state has not itself been consistent in terms of the closing off of early exit routes. Thus though Guillemard and Argoud (2004) argue what they term the 'golden age of early exit' was coming to a close in the early 1990s, they also high-light the persistence of early exit and the tendency of public authorities to resort to these measures in times of high unemployment.

In terms of the actions of social partners, in 1995, the same year as the introduction of the ALMP *contrat d'initiative emploi* (CIE), the com-bined early retirement/hiring scheme (*L'Allocation de Remplacement Pour l'Emploi*, ARPE), also known as *mise à la retraite* was created. This scheme was introduced following an agreement with the social partners[1] involved in the National Unemployment Insurance Administration (*Union Nationale pour l'Emploi dans l'Industrie et le Commerce*, UNEDIC) and was renewed several times until its phasing out was initiated in 2001. The ARPE was introduced in the face of high unemployment as a means to reduce the labour supply (Gendron, 2011). The scheme allowed individuals to exit the labour market early at 58 (55 in special cases) if they had ful-filled at least 160 insurance quarters. These individuals then received an income set at 65% of their former salary and the employer had to recruit an unemployed individual (OECD, 2005e; Taylor, 1998). This scheme was legally separate from dismissal and was only for those individuals who could claim a full pension due to a full contribution record. Drury (1993) argued that this presents a paradox in that workers are protected from dismissal; however a worker's pension record could negate this and Guille-mard and Argoud (2004) argue the ARPE revived the tradition of the 'solidarity contracts'.

Following negotiations with the social partners, new schemes were created at the end of the 1990s. The end of employment leave (*Congé de fin d'activité*, CFA) scheme was established in 1997 for those employed by the state, local public servants and hospital staff. If these individuals were aged 55 and had an extensive contribution record, they could exit and receive 75% of the average gross wage over the previous six months (OECD, 2005e). 1999 saw the creation of a state-provided scheme to allow the early withdrawal at 50 for workers who were exposed to asbestos (CAATA) who received an allowance no higher than 85% of the monthly reference wage (OECD, 2005e). Individuals with an asbestos-related disease could exit at 50 or for those who were not unwell but had worked with asbestos, the exit age corresponded to the number of years they had worked in these conditions (the number of years divided by three would be sub-tracted from the state pension age, e.g. someone who had worked with asbestos for 15 years could exit at 55 years of age). In this year, the job search exemption was also expanded (see *Early exit*) whilst somewhat paradoxically employers' propensity to make older employees redundant was tackled (see *Active labour market policies*).

France's range of early exit measures was reassessed in the early 2000s. The combined early retirement/hiring scheme (*L'Allocation de Remplacement Pour l'Emploi*, ARPE) was beginning to be phased out in 2001 and the ASFNE, which had become increasingly difficult to access, was restricted to only those industries in economic difficulty as of 2000 (France, 2000). However whilst the ASFNE scheme was being closed off, in February 2000 the Early Retirement Scheme for Certain Employees (*Certains Travailleurs Salariés*, CATS) was created by the state following a collective to focus on those employed in arduous or strenuous work (15 years of successive shift work or assembly line work, 200 or more nightshifts a year for 15 years) and in certain industries: automobile, paper and cardboard industry and quarries (Guillemard and Argoud, 2004). The scheme was an extension of the CASA measure to allow older employees in the automobile industry to exit early. Depending upon the agreement of the social partners, under the CATS policy an individual could retire partially or fully at 55. The employer was then exempted from paying social insurance contributions and the state was partially responsible for covering the income of the individual (OECD, 2005e). This scheme, France (2002) argued "is based on a new under-standing of intervention measures for older workers – by handing the responsibility for their implementation back to social partners, by relying on the increased financial involvement of companies, and by encouraging those companies that have recourse to such measures to plan manpower and skills ahead, which should allow them to eventually discontinue their recourse to these early retirement schemes" (France, 2002: 24).

In 2003, a 'National Mobilisation Plan for Employees 55+' was announced, its aim being to discourage companies from using early retirement schemes and older workers from engaging in them. From 27th May 2003, the early retirement benefits that were paid from employer to former employee were subject to an additional tax. Furthermore, the PRP (see *Partial retirement*) and CFA schemes were gradually abolished with these reforms to be completely closed off in 2005. Article 18 of the Pension Reform Law increased the eligibility criteria of the CATS early retirement scheme so as to target older workers who faced multiple disadvantage in terms of the labour market. These reforms also meant that after January 2004, those receiving a pension would be given the opportunity to continue working after the age of 60, providing their new income would be less than their previous wage or than the threshold established according to the geographical area. To complement this, the age at which employers could make retirement compulsory had been extended to 65 from 60. However, employers could be exempted if the redundancy was in line with a collective bargaining agreement or if an early retirement measure was entered into before 22nd August 2003 (France, 2003; OECD, 2005e; Gendron, 2011). Thus the early retirement schemes had been limited to only two main options: for those who had engaged in certain forms of work and those made unemployed through collective social agreements (*plans sociaux*) (OECD, 2005e).

However, while the 2003 Fillon Act altered the *décote* (discount) system, it also introduced a new option for early retirement, albeit fairly restrictive. This scheme, for those with a long career (*départ anticipé pour carrière longue*) comprising of 43.5 years of contributions, allowed individuals to retire prior to 60 if they had begun work aged 15 in 1970. The first entrants would thus be eligible in 2013. The lowest age limit for retiring under this scheme was 56 years of age. In terms of the *décote* (discount) system, the Balladur reform of 1993 had introduced a 10% reduction per year for private sector employees who retired early; the reform of 2003 sought to equalise the private and public sectors by introducing an annual 5% reduction for both by 2005 (Belloni et al., 2006).

The 2007 Social Security Funding Law which was passed on 21st December 2006 complemented the Fillon Act by abolishing the opportunity for employers to make individuals involuntarily retired prior to the age of 65. Between 2010 and 2014, a transition period will allow employers to retire employees with their consent before the age of 65. However, this transitionary period was abolished by the 2008 Social Security Funding law which therefore banned involuntary retirement from 2010. Employers making individuals retire would be required to cover the retirement indemnities at 50% (Bentoumi and Evans, 2008).

The pension reform of 2010 further modified the early retirement route for those with a long career (*départ anticipé pour carrière longue*) by taking into account increased longevity. Thus the scheme was altered from 2011 to include those born in 1960 and began work before the age of 18 with 43.5 years of contributions. The lower age limit was altered to correspond on a sliding scale with the individual's year of birth and when they began work, i.e. certain cohorts (born in 1954–6) could still exit at 56 who began work at or before 16 years of age, whilst those born in 1960 could exit at 58.

Deferred retirement

In the early 1990s, individuals could increase their pension amount from the age of 65 by 2.5% per year if 150 insurance quarters had not been reached. Those who had reached the minimum contribution requirement could not increase their pension or combine work and retirement (Ebbinghaus, 2006). The 2003 pension reforms introduced a *surcote* (premium) to encourage flexibility around the state pension age, and to complement the *décote* (discount) measure. In terms of the former, individuals who had fulfilled the requisite insurance quarters could retire at 60 with a full pension, or, could continue to work and receive an additional pension amount (0.75% per quarter or 3% per year) (MISSOC, 2006; Social Security Administration, 2008; Belloni et al., 2006). This had increased by 2011 to add an additional 5% for individuals who had made the requisite 41 years for a full pension who continued working beyond the age of 60 (OECD, 2011).

Labour market policies

Early exit

In the peak period of early exit policy in the 1970s, France introduced two measures, the GRL (*Garantie de Resources Licenciement*) and the GRD (*Garantie de Resources Demission*) in 1972 which were disbanded in 1983 as the cost sent the unemployment insurance scheme operated by the social partners (UNEDIC) into financial crisis (Mandin, 2004). However, as of 1984, those over the age of 55 were exempted from seeking work and signing up for a 'return-to-work action plan' (*Plan d'aide au retour à l'emploi,* PARE) under a scheme called *Dispense de Recherche d'Emploi* (DRE). In 1985, this was altered to cover those aged 57 and over. The Association for Employment in Industry and Trade (*Association pour l'emploi dans l'industrie et le commerce,* ASSEDIC) would send those eligible for the DRE a request form which would result in their removal from the unemployment register (OECD, 2005e). In terms of the unemployment benefits available in 1995, the durations were contingent on age and contributions. In addition, those over the age of 60 could only receive unemployment insurance (*assurance chômage*) if they had not made the requisite insurance quarters (151) in order to receive a full pension amount.

In the same year as a new active labour market policy was introduced (the consolidated employment contract (*contrat emploi consolidé*), the ASA (*Allocation Spécifique d'Attente*) was created in 1997 for older unemployed individuals who had exhausted their right to unemployment benefits or minimum income support allocations (*Revenu Minimum d'Insertion,* RMI), who were over 60 years of age and had 40 contribution years (Guillemard and Argoud, 2004; Jolivet, 2002). Once again, the French reforms on the one hand moved towards active ageing whilst at the same time, not only preserving the right to exit early, but extending it. In addition, the 1997 Older Unemployed Compensation (*Allocation Chômeurs Âgés,* ACA) provided an additional unemployment benefit duration for those who had made 40 years of contributions from the age of 50 (thus though the ASFNE age threshold had been raised, this route provided an alternative) (Guillemard and Argoud, 2004). As aforementioned, this was a period of high unemployment and thus this scheme was part of a tradition in France whereby early exit was used to reduce the labour supply (Gendron, 2011).

In 1999, reform was undertaken regarding the requirement to be actively seeking work, the *Dispense de Recherche d'Emploi* (DRE) scheme. This scheme was expanded in 1999 to allow those with 160 contribution quarters (40 years) to exit at 55 (the general rule was 57) (OECD, 2005e). Though in 2002 PAP-ND project (Personalised Action Plan for a New Start) focused attention on older unemployed individuals (see *Active Labour Market Policies*), in this year, a new means-tested allowance was introduced. The *Allocation équivalent retraite* (AER) was for unemployed individuals over the age of 60 who had made 40 years of contributions, which replaced the ACA (*Allocation Chômeurs Âgés*) and the ASA (*Allocation Spécifique d'Attente*) (Bertelsmann-Stiftung Foundation, 2004c; France, 2002). In addition, the Specific Assistance Allowance (*Allocation de solidarité spécifique,* ASS) was created for those over the age of 55 which provided an unemployment benefit supplement if they had worked for five out of the last ten years. An individual could combine work and training whilst still receiving the allowance for the first few months (OECD, 2005e).

Active labour market policies

In order to protect the employment rights of older individuals, from 1987 any employer terminating the contract of a worker over the age of 50 had to pay the unemployment administration (UNEDIC) a contribution called *Delalande*. The *contrat de retour à l'emploi* (CRE) was introduced in 1991 which provided financial incentives for employers to hire older workers (Gourin, 1998).

In terms of active labour market policies, in 1995 *contrat d'initiative emploi* (CIE) replaced the *contrat de retour à l'emploi* (CRE) to provide funds for employing older people at the end of their career and encourage the employment of the long-term unemployed (Gourin, 1998). The CIE scheme provided employers with a subsidy of around 2,000 francs a month and removed the social security contributions the employer was required to pay for 24 months or permanently if the individual was disabled or unemployed and on social assistance for more than one year (OECD, 2005e, 2006).

New measures were introduced to encourage older workers' employment. Those over the age of 50 could also engage in a consolidated employment contract (*contrat emploi consolidé*). This scheme was introduced in 1998 and provided a fixed-term 12-month contract for those with limited employment prospects. The individual had to work a minimum of 30 hours a week and the state covered the income up to 120% of the national minimum wage. Again, the employer was exempted from social security contributions (OECD, 2005e).

Though 1999 also saw the expansion of the job search exemption and the creation of a new early retirement route (CAATA, see *Early retirement*), this year did see one reform to combat early exit with the *Delalande* contribution doubled in 1999 for companies with 50 or more employees. The contribution was increased to be equivalent to two months wages for employees aged 50, rising with age to be equivalent to one years' salary at 57 years of age (France, 2002).

2001 saw the creation of anti-ageism legislation to cover recruitment, promotion, training and dismissal in the form of a Labour Code on age discrimination. However, exemptions applied if the discrimination had a 'legitimate objective' (OECD, 2005e). On 16th November 2001, new clauses were added to anti-discrimination legislation. The burden of proof was placed on the employee and trade unions were permitted to speak in their stead during juridical proceedings (France, 2002). This year also saw the establishment of Personalised Action Plan (*Plan d'action personnalisé*, PAP) to provide unemployed individuals with training, help with job search and skills assessments (OECD, 2005e). As of 2002, the PAP-ND project (Personalised Action Plan for a New Start) sharpened its focus on those over 50 who had been unemployed for a substantial period (France, 2002).

2003 marked a sea-change in the French policy towards older individuals. Not only had a number of the early retirement routes been curtailed, a number of ALMPs were introduced. In 2003, the Degressive Employer's Subsidy (*Aide dé à l'employeur*, ADE) was reformed to allow those over the age of 50 to access it from the third month of unemployment as opposed to the previous threshold of one year. The state partially covered this subsidy for up to three years (OECD, 2005e; France, 2003). The *Delalande* tax for employers was reduced so as to prevent it becoming an obstacle to the hiring of employees over the age of 45 (Gendron, 2011). The rules around combining work and the solidarity allowance were also relaxed so as to encourage those reluctant to enter employment due to the precarious financial situation created by the move from benefits to work. To make the recruitment of older workers with additional labour market barriers more attractive to employers, the CIE subsidy was extended from two to five years and would be paid quarterly (as opposed to annually). In addition, the required unemployment period was reduced from 24 to 18 months.

In 2003, lifelong learning measures were also introduced for those over 45 years of age with 20 years of experience. This scheme, *validation des acquis de l'expérience* (VAE) provided these individuals with a skills audit, followed potentially by access to work experience, and training was made more accessible (France, 2003; Taylor, 2005). December 5th 2003 saw the creation of the multi-sector lifelong learning agreement signed by all the union organisations. This agreement aimed to make lifelong learning practices better equipped to deal with the needs of older individuals. As a result, from 2004 companies with more than ten employees contributed 1.6% of the annual payroll to training for employees; an increase of 0.1%. Those firms with less than ten employees were able to make reduced contributions of 0.4% (as opposed to 0.5%). This scheme entitled all full-time employees to at least 20 hours of vocational training per year, known as the 'individual right to training' or *Droit Individuel à la Formation* (DIF). The DIF could be accumulated over a six year period but if it was not wholly or partially used by this time, it was limited to 120 hours.

Employees could transfer this entitlement when they moved employers and those who worked part-time could receive a proportional amount of training hours. The employee had autonomy over the training yet required the consent of their employer (France, 2003).

France implemented age discrimination legislation in 2004 which covered employment. The 2004 legislation prohibited age limits yet mandatory retirement at the age of 65 was still permitted and companies that could provide justification could dismiss workers prior to this threshold (OECD, 2006). In January 2005, the High Authority to Combat Discrimination and Promote Equality (HALDE) was established to process cases and disseminate examples of best practice (OECD, 2005e; Bertelsmann-Stiftung Foundation, 2003d). On 6 June 2006, a new plan to promote the employment of older workers (*plan emploi seniors*) was announced and included the *Contrat Dernière Embauche* or CDD, which was a 18-month temporary employment contract for those aged over 57 and unemployed for three months or more. This plan also included €5 million information campaign to promote positive images of older people and lifelong learning (Milner, 2007). However, following the National Action Plan of 2005–10, France phased out the *Delalande* from 2005–10 which had afforded older workers some additional protection from redundancy (Guillmard and Jolivet, 2008).

Part-time work/pension arrangements

In 1982 the solidarity gradual preretirement contracts (ACC) allowed those aged between 55 and 60 to partially retire from the labour market. The firm was required to sign an agreement regarding the replacement labour and in return would receive state subsidies[2] (Guillemard, 1991; Ebbinghaus, 2006). In the same year, the phased-in retirement (*cessation progressive d'activité*, CPA) was created for civil servants aged 55 and over with 25 years of service. Under this scheme, the individual could reduce their working hours to 50% for which they would receive 50% of their wages in addition to a 30% supplement (Jolivet, 2002; Guillemard, 1991; Eichhorst and Rhein, 2005).

In 1988, the gradual retirement scheme (*retraite progressive*) was established which allowed the individual to claim part of a pension and work part-time from the age of 60 as a means to extend working lives (Ebbinghaus, 2006; Jolivet, 2002). In an attempt to encourage gradual retirement for those in the public or voluntary sector, the employment solidarity contract (*Contrat emploi solidarité*, CES), modified in 1990, permitted part-time employment for those over 50 years of age for a maximum of two years. The state covered between 65 to 80% of the wage and the employer was exempted from social security contributions (OECD, 2005e).

In 1993 the phased early retirement scheme (*préretraite progressive* (PRP)) replaced the ACC scheme and required the employer and state to form an agreement which would allow employees over the age of 55 to enter part-time employment and receive an allowance of 30% funded by the employer and state. The employer then had to replace the worker with an unemployed individual. This scheme and the ARPE continued the tradition of the 'solidarity contracts' by combining "employment (early exit as a way to free jobs for young people in particular) and social justice (early exit as a compensation for wage-earners who have paid into the Old-Age Fund for too long)" (Guillemard and Argoud, 2004: 172).

In 2005 a new phased retirement scheme was implemented to replace the PRP scheme and to promote free choice for individuals. Those individuals who had made 33 years of contributions could enter into the gradual retirement scheme and continue working, providing their salary and part-time pension income when combined was not higher than their previous wages (France, 2004). These employees could withdraw a percentage of their pension whilst remaining in work (OECD, 2006). The employment solidarity contract (*Contrat emploi solidarité*, CES) was also replaced in 2005 with the Accompanying Contracts of Employment (*Contrats d'accompagnement dans l'emploi*, CAE). This scheme had many of the characteristics of its predecessor, in addition to providing a subsidy for between 12 and 24 months, the individual could also receive training for between 200 and 1,000 hours in order to meet the job requirements. The individual had to have been unemployed for 12 of the previous 18 months. The programme included those between the age of 50 and 65 but did not apply exclusively to this group (Ministère du Travail, des Relations sociales, de la Famille, de la Solidarité et de la Ville, 2005).

7.4   Italy

In Italy, the employment rate for older workers was the lowest of all EU15 nations at 36.6%, significantly below both the Stockholm target and the average rise of all nations over the period of 2001–10 (+9.6%) (Eurostat, 2011). Ageing and employment did not appear on the policy agenda in Italy and the problem of unemployment was underestimated due to the economic growth in the late 1980s (Feroldi, 1998). As a result of this and the availability of early retirement and exit, between 1960 and 1990, the effective retirement age for men fell by five years to 59 and from 62 to 57 for women. The situation of older workers was compounded by limited school enrolment until the 1960s. Consequently, older individuals have a lower level of education than their younger counterparts with 60.3% of those aged 55 to 64 in 1998 having only obtained an elementary school leaving certificate (Italy, 1999). In addition, there were regional disparities between the North and the South with the latter exhibiting high youth and female unemployment. However, the 1990s saw a change with concerns over welfare costs accompanied by a reduction in the pre-retirement schemes in order to increase the effective retirement age (OECD, 2005h).

Pension policies

Pension principles and state pension age

The state-provided Italian pension system comprised of several pillars in the early 1990s. The first and primary pillar was deferred according to occupations on the basis of contributions. The second pillar therefore provided for those individuals with insufficient contribution records on a means-tested basis (Ferrera and Jessoula, 2007). In addition, a third and fourth pillar were added, which were voluntary and subsidised by the state. For individuals insured before 1993, there was also the *Tfr* scheme which provided a defined benefit from employers to employees in the event of redundancy (Ferrera and Jessoula, 2007). State pension ages were 55 for women and 60 for men in 1990, with an employment history of 15 years (MISSOC, 1994).

In the 1990s, there were three main years of reforms that tackled the issue of demographic ageing. The *Amato* (law 503/92) reform in 1992 gradually increased the state pension age from 60 to 65 for men and from 55 to 60 for women by 2002 (OECD, 2005h). The reference earnings period for pensions was also changed from the previous ten years and the indexation based was shifted from wages to prices (Ferrera and Jessoula, 2007). The minimum contribution requirement was also raised to 20 years. However, individuals with over 15 years of contributions in 1992 were exempted from these new rules, with the exception of the change to pension indexing whilst those with less than 15 years were subject to the new and old systems in proportion to the years of contributions pre- and post-1995, whilst those entering the system as of 1992 would be entirely subject to the *Amato* rules (OECD, 2005h; Billari and Galasso, 2008).

The 1995 *Dini* reform took the *Amato* reform further to attempt to maintain the pension system's sustainability as well as incentivising labour market participation (OECD, 2005h). Law No. 335/95 on the

Reform of the Pension System introduced a number of new measures. By this year, the state pension ages had been increased to 62 for men and 57 for women in line with the 1992 *Amato* reform. Under the *Amato* reform, the state pension age was increasing by one year every two years that passed; the *Dini* reform raised this to two years every 18 months (MISSOC, 1996). The *Dini* reform however "largely protected the 'acquired rights' of older workers by introducing long transition periods" (Ferrera and Jessoula, 2007: 437). Thus, as with the *Dini* reform, three groups were identified: those who had made 18 years of contributions prior to 1995 were subject to the rules of the *Amato* reform; those with less than 18 years of contributions would partially receive their pension in accordance with the old and new systems in proportion to the years of contributions pre- and post-1995; and finally, those entering the system after 1995 would be subject to the new rules. As a result, this new system would be fully in place by 2030–5 (Ferrera and Jessoula, 2007).

The rules regarding the phasing in of the *Amato* and *Dini* reforms produced significantly different pension structures for individuals who were in all respects very similar but with a one year difference in their contribution record (Thompson, 2009). Billari and Galasso (2008) use the example of two workers in the private sector who entered the labour market at 20, but were born on year apart, in 1957 and 1958. By 1992 when the *Amato* reform was introduced, the person born in 1957 had made 15 years of contributions and was thus subject to the pension system regulations made prior to this ruling. The individual born in 1958 had however made 14 years of contributions and therefore their pension amount would be calculated on a pro-rata basis of almost two-thirds (26/40) in accordance with the *Amato* rules according to the new rules and the remaining part (14/40) in line with the pre-*Amato* reform scheme. If they both opted to retire at 60 with 40 years of contributions, the former person born in 1957 would be able to retire at 2017 with a pension benefit at 80% replacement rate whilst the individual born in 1958 would be able to retire with a replacement rate at 70% due to the application of a mix of pre- and post-*Amato* rules. In addition, when the pension ruling introduced by the *Dini* reform also applied to the individual born in 1958 – as they would not have achieved 18 years of contributions by 1995 either – their pension would be reduced further to 65% of the replacement rate. Assuming the individual born in 1957 continued to work between 1992 and 1995, they would have made the 18 years of contribution necessary to receive a pension in line with the pre-*Dini* system and they would therefore have still received a pension at 80% of the replacement rate.

The key change introduced by the *Dini* reform was to move away from a defined benefit scheme to a 'notional' defined contribution system. Thus the pension payment would be directly linked to the contributions made over an individual's lifetime (OECD, 2005h). Under the new system introduced by the *Dini* reform, a flexible pension age was introduced between the ages of 57 and 65. The minimum period of contributions was five years and the pension accrued had to be 1.2 times the amount of the social allowance in order for the individual to exit. Though pensions could be accessed as early as 57, there was a strong disincentive as the pension amount would increase with age and contribution years (OECD, 2005h; Belloni et al., 2006).

Post-1997, there had been several increases in the minimum amount of social assistance pensions and incentives to encourage people to work beyond the minimum threshold for state pension receipt. In 1997, the financial Law number 303/96 (subsection 185) encouraged individuals to retire gradually. By 2000, the state pension ages had been raised to 60 for women and 65 for men for those under the *Amato* (pre-*Dini*) pension rules i.e. those introduced in 1992 for which those with more than 15 years of contributions at this point were exempted.

Early retirement

In Italy in the 1990s, in terms of early retirement, employees working in industries facing economic difficulties could also retire prior to the state pension age. This scheme was initially opened in 1968 and closed in 1979 due to the poor take-up as a result of the low benefit levels. It was reopened again as the *'prepensionamento'* scheme, financed by the special unemployment fund initially before employer co-payments were introduced. Workers made redundant from industries in financial difficulty with 15 years of contributions would be able to receive pre-retirement benefits (for men from 55 and women from 50 unless they were employed in the steel and shipping industries where the age thresholds were 52 and 50 respectively) (Ebbinghaus, 2006).

In addition, there was also the Seniority Pension (*pensione di anzianità*) which required a certain level of contributions to allow for early retirement. These contribution requirements prior to the 1992 reforms differed according to sector with public sector workers required to have made 20 years of contributions (15 in some cases) to receive the Seniority Pension, whilst private and the self-employed needed 35 insurance years. In 1992, the requirements for the public sector were harmonised in line with private sector employees and the self-employed. Thus individuals could retire at any age under the aforementioned Seniority Pension, providing they had made 35 years of contributions though delaying pension receipt for one year would result in an increase of 5.7% and 6.5% for two years (Mirabile, 2004; Paulli and Tagliabue, 2002). As a result, this scheme allowed individuals who entered the labour market at 15 to exit at 50 years of age (MISSOC, 1994). The Seniority Pension became problematic when the post-war generation began to retire: the numbers claiming rose from 289,000 in 1990 to 865,000 in 1996 (SZW, 1997 in Ebbinghaus, 2006).

As aforementioned, under the *Dini* system, though pensions could be accessed early due to the flexible pension age for certain birth cohorts, there was a strong disincentive as the pension amount would increase with age and contribution years (OECD, 2005h). The 1995 reforms also altered the Seniority Pension to create two options: to retire at any age with a certain level of contributions and to retire at a set age with a certain (slightly lower) level of contributions. These changes would be actualised to gradually reach 40 years by 2008. The *Dini* reforms were deemed not to have gone far enough to stabilise pension spending and so in November 1997 the *Prodi* (449/97) reforms were introduced. These reforms accelerated the changes to contribution and age thresholds for the Seniority Pension to restrict access. As a result, as of 1998, to exit at any age, the minimum requirement had been raised to 36 years for both public and private sector employees and 40 years for self-employed individuals; early retirement was also possible at 53 years of age if employed in the public sector, 54 if employed in the private sector and 57 if self-employed with 35 years of contributions. From 1998, these thresholds would continue to rise at different rates to reach 40 contribution years to retire at any age and 57 with 35 years of contributions unless they were self-employed when the latter option was available from the age of 58. Slightly lower age thresholds applied to those on CIGs (see *Early exit*), blue-collar workers, those in 'precarious' or risky employment and those who had been employed before 19 years of age; the age threshold of 57 would not have applied to these groups until 2006 whereas for the private and public sectors, this would have been in place as of 2002 and 2004 respectively (Inglese, 2003; OECD, 2005h; MISSOC, 1999).

On July 28th 2004 a law was passed in the Italian parliament to further reform the pension system. The main changes included the establishment of a retirement age for the Seniority pension to 60 as of 2008, gradually rising to 61 by 2010, and to 62 by 2014 for men with 35 years of contributions. Women will be able to access this pension from age 57 until 2015 with 35 years of contributions; as of 2016, women will not be able to retire prior to 60 years of age (Italy, 2004; OECD, 2007). However, individuals could continue to retire at any age with 40 years of contributions (Ferrera and Jessoula, 2005).

Deferred retirement

In 1995, deferral of pension receipt would result in an increase to the state pension amount by 3–3.5% per year. Further incentives to delay retirement were reformed in the Budget Law of 2001 and made effective in 2002. To encourage individuals eligible for the Seniority Pension (*pensione di anzianità*) route to remain in the labour market, they were able to postpone exit for a minimum of two years and be exempted from their social security contributions, thereby increasing their income by somewhere in the region of 50% (Bertelsmann-Stiftung Foundation, 2002a). In addition, the pension rights would continue to be accrued and this also applied to those working beyond the state pension ages (65 for men and 60 for women) (Mirabile, 2004).

In July 2004, the Italian parliament passed a law to reform the social security system. This law, the NAP of that year argues, aimed to "a) to gradually increase retirement age in order to take into account demographic trends; b) to develop complementary pension and insurance schemes alongside the state system so as to ensure better system sustainability" (Italy, 2004: 18). It was decided that financial incentives should be implemented from 2004 to 2007 for individuals in the private sector to encourage them to continue working after they had accrued sufficient pension credits. Should the individual decide to continue working, they would receive an increase to their wage equivalent to the social security contributions they would have been paying; an increase which could be between 32.7% and 45% of their total salary and this amount would be exempted from taxation (Italy, 2004).

Labour market policies

Early exit

Despite the raising of retirement ages in 1992, measures were introduced in 1991 to allow for exit as early as 55 years of age, thus representing somewhat of a paradox in terms of the direction of reform. First, the *cassa integrazione guadagni straordinaria* (wage guarantee fund for those made redundant, CIGs) Law 233/91 provided workers made redundant in times of economic crisis with a benefit of 80% of their previous wages who were then placed on 'Mobility Lists' whereby new employers could receive a subsidy if they offered them work. In the case of older workers, the CIGs scheme soon came to provide a 'bridge' to retirement due to the additional benefit duration it provided for those over the age of 49 (Ebbinghaus, 2006; Paggiaro and Trivellato, 2002). Whilst those aged under 40 could receive this benefit for one year and those aged 40–49 could receive it for two, individuals over this age could receive CIGs for three years (Paggiaro and Trivellato, 2002). In addition, for those with 15 years of contributions and five years remaining until state pension age, this could be extended further until retirement under 'Long Mobility' (Dell'Aringa and Samek Lodovici, 1996).

Active labour market policies

The expansion of early exit policies of the early 1990s conflicts somewhat with other elements of Italian labour law. In accordance with Article 3 of the Italian Constitution, there is a general principle of equality for all citizens. This was further strengthened with Article 15 of Law 300/1970 which prohibited discrimination on the grounds of race, religion or gender to which age is considered a natural extension (OECD, 2005h). In Italy the constitution allowed for the choice to work, with no discrimination on the basis of age. Work in Italy was thus a right and the constitutional court ruled that there should be legislation to ensure equal rights for older people. Older workers were protected in industry by a 'first in, last out' policy of redundancy, yet were still over-represented in periods of downsizing. In terms of government-provided training measures, Italy exhibited direct discrimination by setting upper age limits (Drury, 1993). Some positive measures were introduced, including a working group on the issue as well as lifelong learning schemes with funding from the European Social Fund. Regional governments have also introduced financial incentives for the recruitment of the 50-plus, support for community projects that employ this age group and funds for employers with training and mentoring for this group in place.

The *Biagi Law* (Legislative Decree 276 October 2003) focused on age discrimination and created a scheme targeted at older individuals. The first Article of this law simplified the process of job placement in the private and public sectors. Article 2 promoted lifelong learning and was the first attempt to address education for older workers since Law 236/93. Article 3 sought to protect part-time workers and non-standard work such as for people on-call was covered by Article 4. Article 5 supported collaboration between employers and trade union representatives (OECD, 2005h). This decree also included individuals over the age of 50 under the heading of 'disadvantaged workers' and could thus utilise the social employment or personal services agencies and the incentives they provided. In addition, this law created placement contracts that were non-renewable and lasted between nine and 36 months (Ebbinghaus, 2006) and flexible contracts or opportunities for self-employment. There was also the option for those over 50 who had been unemployed for two years to engage in projects designed to update skills (Italy, 2004).

Part-time work/pension arrangements

With regard to partial retirement, Law no. 335/95 of 1995 meant that workers two years under the old age pension age could combine work and pension receipt for this period, providing they had made 37 years of contributions (Mirabile, 2004; Paulli and Tagliabue, 2002). The partial pension route was modified as part of the Budget Law of 2001 and made effective in 2002. After this point, the scheme applied to those who had made 40 years of contributions and had fulfilled the eligibility for the seniority pension (35 years of contributions) before 1995 (Paulli and Tagliabue, 2002).

7.5   Summary and impact at the individual level

Spain, Greece, Italy and France share commonality in terms of their wide range of early exit and retirement schemes available in the early 1990s, and their comparative belatedness in terms of retrenchment. These nations tended to undertake the reforms of these policies in the early- or mid-2000s, whilst consolidating or even expanding the policies on offer for exit in the 1990s. In terms of ALMPs, Greece focused on encouraging employers to recruit older individuals through subsidies and provided training measures; Italy, France and Spain too increased incentives for employers and boosted the protection offered to older workers through anti-ageism legislation and training measures. In addition, Greece could see a reduction in early exit in the future due to their differentiation between cohorts, with less opportunity for pre-retirement available to younger generations.

Spain in the mid-1990s partially focused on the labour market in terms of its decommodification policies, thus providing equal opportunities for all the model biographies made redundant or employed in an industry in crisis to retire early. They all would also have been exempted from the job search requirement of unemployment benefit receipt. In addition, *Sasha* (with 35 contribution years and aged 63) would have been able to opt to exit at 60 due to their birth cohort and contribution record, though this would have resulted in a reduced pension amount. There was also the option for *Laurent* (55 with 35 years of employment contributions) and *Sasha* (63 with 35 years of employment contributions) to exit early if unemployed through a benefit extension due to their contribution record, representing a desert focus. ALMPs were introduced over the following five years, including those aimed at the over-60s whilst the decommodification options were largely unreformed. Both *Sasha* and *Jude* could combine work and pension receipt.

By 2010, there was increased compulsion to engage in job search. However, the early retirement schemes for certain forms of work were retained, and the age threshold for the *'Relevo'* (take-over) scheme was lowered to 60 and the scheme for those with long insurance durations was expanded to those who had made 30 years of contributions, not just those who had 15 insurance years before 1967. Thus Spain strengthened the desert focus of its decommodification policies, whilst the labour market elements remained unretrenched. The partial pension scheme had become more restrictive, only available to those with 30 years of contributions.

Italy retained its early retirement eligibility focus on a mix of labour market needs and deservingness with a rights-focused early exit option over the 15 year period, as reflected in the policy treatment of the model biographies. The state pension ages in Italy in 1995 meant that *Sasha* would have been able to exit at 60 if male and 55 if female as they were subject to the state pension ages applicable before the *Amato* reform due to their contribution records pre-1995, though they would have been able to continue to work until they had reached the 40 contribution years required for a full state pension. They would also have increased their pension by 3–35% per year they worked post-state pension age. Similarly, if *Laurent* had been female, she could have exited at 55 under the conditions of the pre-*Amato* reform pension system. Had they been male, there were exit options available in the mid-1990s including the Seniority Pension, the CIGs unemployment scheme with an extension for older individuals, the *'prepensionamento'* scheme for industries in economic difficulty, and the 'Long Mobility' option which would have enabled exit until the state pension age. *Jean* too would have been able to receive the CIGs benefit for three years but the other options were not available due to their contribution requirements. The lower state pension age for women was not applicable to *Jean* as they had not made sufficient contributions pre-1992. *Jude's* (aged 63) disjointed work history would have meant they would have been subject to a mix of the newer, flexible *Dini* system and the *Amato* system, allowing them to retire between the ages of 57 and 65. However, the pensionable income produced by this system was contingent on contribution years and thus *Jude* may have been more likely to attempt to remain in the labour market for as long as possible. *Jean* too would have been able to receive the CIGs benefit for three years but the other options were not available due to their contribution requirements.

Over the following five years, the 1998 *Prodi* reforms accelerated the increases to the Seniority Pension's age and contribution requirements. As a result, as of 2003, a male *Laurent* would no longer have been able to access this option as the age threshold had been raised to 56 and the contribution requirement for the option without an age limit had also increased to 37 years. However, had he been employed in blue-collar work, in 'risky' employment, had begun his career between the ages of 14 and 18 or been enrolled on a CIGs, he could have accessed the Seniority Pension until 2005 when the age threshold was raised. The bonus for deferring state pension receipt had improved in 2004. However, unlike many of the EU15 nations, Italy was slow to adopt ALMPs and between 1996–2000 had made limited progress in this regard. Indeed, in the period of 2001–05 one of the incentives for those over 60 to remain in the labour market through the exemption of social security contributions contained a contribution requirement, thus barring *Jude*. By 2010, Italy had introduced ALMPs, yet had not retrenched the CIGs early exit route available to all of the model biographies. However, *Laurent* would no longer have been able to access the Seniority Pension in instances of extensive contributions as the age threshold had been raised for those with 35 insurance years.

In the case of France, age was a key factor in decommodification options. *Sasha* and *Jude* would have been able to exit at 60 years of age in line with the state pension age over the entire period, which was comparatively low among EU15 nations. *Sasha* and *Jude* could continue to work and increase their pension amount, but only as they had not made sufficient contributions to receive a full pension (37.5 contribution years). With regard to early retirement *and* exit, the decommodification options largely contained a mix of desert and labour market considerations in terms of their eligibility criteria in 1995. As a result, *Jean* would have been barred as though they may have worked in particular industries, they had not made the required insurance contributions. *Laurent* and *Jean* would have been able to combine work and pension receipt through the PRP (*prére-traite progressive*) and CES (*Contrat emploi solidarité*) schemes respectively.

However whilst the following nine years saw the expansion of ALMPs, early exit and retirement routes were also added, which could be seen to represent a contradiction. Yet these options for decommodification were only applicable to certain industries so as to restructure firms or because these occupations were particularly arduous, demonstrating France's commitment to decommodification as a means to assist the market and perhaps a testament to the power of the various occupations' unions in securing early retirement for their members. *Jean* again would have been barred as although the new options for those in arduous work (CATS) or in road transport (CFA), the retirement options also contained a requirement for careers of a certain length (the exception would be if they were suffering from an asbestos-related disease in which case they could have accessed the CAATA scheme at 50).

The Greek system also provided unharmonised age thresholds for men and women, as well as different birth cohorts throughout the entire ten year period; thus for those insured prior to 1993, the state pension ages were 60 for women and 65 for men whilst for those who entered the system after this point, the age for both genders was 65. Initially, the early retirement options were a mix of those based on occupations and contributions, and thus if *Laurent* and *Jean* were female and employed in certain professions, they could both have exited early with no reductions to their pensions. In addition, there was also the option of exiting early if *Laurent* and *Jean* were female with a reduction of 6% per year prior to the state pension age (30% for these individuals). For *Sasha* and *Jude* in 1995, retirement at 60 under the regular pension was available to them if female due to the unharmonised state pension ages. In addition, exit would have been available if they were male under similar conditions as *Laurent* and *Jean,* i.e. if they worked in arduous labour or the construction industry or at universally at 60 with a 30% pension reduction. However, *Sasha* would also have been able to access the route for long insurance durations with no reduction. Both *Laurent* and *Sasha* would have been able to exit via an extended unemployment benefit duration due to their contribution records and yet there was also the option of subsidised employment (the presence of which in 1995 was comparatively early among EU15 nations).

Within five years, the early retirement routes for arduous labour and the open scheme with reductions had contribution requirements added, thus barring *Jean* and *Jude.* In addition, the option of retiring earlier under the schemes previously accessible to *Laurent* and *Sasha* were also replicated for younger ages, but with reductions added, thus delaying for a few years would still have enabled people retire early without pension reductions. A new special scheme was added for women caring for dependent children who also had certain levels of contributions. This presents a contradiction: the increased care burden of women with children who are minors or disabled is acknowledged and exit is permitted to undertake this care yet at the same time, these individuals may find the contribution threshold more difficult to attain. Concurrently, ALMPs had been introduced to promote the labour market opportunities of these individuals. By 2010, a further desert-focused route was added requiring 37 years of contributions, thus heightening the deservingness-focused criteria of the Greek decommodification system and reducing the options available to *Jean* and *Jude* (and the sub-groups they represent) to exit early from the labour market.

8
Policy Convergence, Divergence and Intragenerational Equity in EU15 Nations

This chapter will bring both the macro and micro empirical elements explored in Chapters 4–7 together for an analytical discussion with reference to the literature from the first two chapters. Following the argument that early exit and retirement policies allow for the decommodification of labour, the retrenchment of these policies therefore represents a *de facto* shift towards recommodification, whilst the introduction of active labour market policies for older workers represent the *de jure* recommodification of labour. As explored in Chapters 2 and 3, the EU advocates the narrowing of decommodification options and the expansion of policies for the recommodification of labour as part of its 'active ageing' agenda. This strategy, it is argued, is necessary to achieve the end of increased and prolonged labour market participation in older age and in turn prevent intergenerational conflict and maintain welfare arrangements. Thus where nations have retrenched their policies for the decommodification of labour whilst at the same time expanding their policies for the recommodification of labour, they could in the context of the EU targets and guidelines be considered to adhere to this particular organisation's active ageing agenda. This chapter therefore addresses three questions: first, are all nations converging towards the EU-vision of active ageing? Second what was the character of reforms undertaken in these nations over this time period? Once these have been addressed, thirdly, the differences at the micro-level will be explored in terms of the impact of gender, age and labour market opportunities on the policy options available to older individuals in EU15 nations.

From the data on EU15 nations' policy reforms over the period of the mid-1990s to 2010, the sweeping generalisation could be made that there has been progress towards this active ageing agenda, which in turn can be seen as indicative of the general move away from decommodifying social policies towards a more recommodifying welfare approach. There has been a broad shift towards the closure of *de facto* and *de jure* early exit routes, the creation of ALMPs that focus on older individuals and incentives for the deferral of state pension receipt. Thus the main onus of the welfare state is no longer to provide security outside the labour market, rather it is the responsibility of the individual to secure their own welfare *within* the labour market and the state's role is to enable this.

However, the idea that there is a universal shift across EU15 nations towards active ageing and the recommodification of older individuals' labour is over simplistic both at the macro- and micro-level. In terms of the former, nations are progressing towards the dual active ageing aims of re-integrating and retaining older individuals in the labour market with different policy mixes and at different speeds, and from different starting points, thus reflecting the idea that policy legacies influence reform paths (Pierson, 1996, 2004). The picture is therefore more complex than linear convergence towards the EU's active ageing approach, representing the recommodification of labour. However, change has occurred, perhaps to a greater degree than the new institutionalist literature would envisage without the necessary 'critical juncture' (Pierson, 2004). Similarly, at the micro-level, individuals within the category of 'older' are experiencing the active ageing agenda differently in terms of the policy options available to them.

Thus the data indicate that though nations are adopting policies that move towards the EU's goals related to active ageing (to increase the numbers of older people in the labour market, and to extend working lives), there is variation both at the macro- and micro-levels and these distinctions will be addressed in this chapter in relation to the two EU policy strands: first retaining older workers in employment, and second re-integrating older individuals into the labour market (which reflect the EU's Barcelona and Stockholm targets respectively). With regard to retaining older individuals in employment, the reform and retrenchment of early exit and early retirement policies are considered as well as incentives for deferring state pension receipt, whilst ALMPs reflect the aim of re-integrating older individuals in the labour force.

8.1   Policy convergence and divergence: The macro-level

This section will explore the two foci of EU active ageing policy separately – the retention of older workers in the labour market and the re-integration of older unemployed individuals into employment – before addressing EU15 nations' overall active ageing policy approaches, including both pension and labour market policies.

Retention: Convergence and divergence in policy contexts and reform

The first area of deviation *vis-à-vis* active ageing and the de- and recommodification of older individuals' labour relates to the original policy contexts in EU15 nations. Nations had in place different decommodification schemes that ran counter to the EU's active ageing goals by allowing individuals to retire or exit early from the labour market which affected both the distance nations had to travel towards the active ageing agenda and the direction the reforms or retrenchment took. In the mid-1990s, the active ageing agenda had yet to be launched by the EU and embraced by EU15 nations. 1995 is the year that EU15 was formed through the addition of Austria, Finland and Sweden and thus this makes a suitable juncture to consider unemployment and pension policy developments from. The legacy of the economic difficulties of the 1970s and 1980s meant the policy focus was still on the decommodification of older workers in many countries. The national policy approaches to older individuals in 1995 can be divided into those where decommodification was bestowed as a right for all individuals over a certain age, to those deemed deserving as a result of their contributions, or for labour market ends (i.e. for certain industries, for firms in economic decline or restructuring, with replacement requirements). What became clear from the data is that classification was made more complex by the multitude of policies available in some nations. Indeed, both extended benefit durations and job search exemptions for older people represented *de facto* early exit, whilst early retirement also allowed for decommodification; in some nations when considered in conjunction, these policies represented an area of contradiction or placed the country within more than one category.[1]

In terms of the nations with rights-based decommodification approaches in 1995, Ireland offered early retirement to all individuals over the age of 55 who were unemployed for a year under the PRETA scheme. Sweden too offered age-related unemployment benefit extensions and all individuals were able to retire early and receive the same penalty per year prior to the state pension age in 1995. There were also relaxed conditions for the receipt of Disability Insurance for those over 60. Conversely, Austria largely focused on desert with regard to the decommodification policies available in 1995. With the exception of a job search exemption and an early retirement scheme in the instance of a work accident, all of Austria's policies contained a contribution requirement, including the two occupation-related options (for unemployed miners (*Sonderunterstützung Bergbau*) and those engaged in strenuous night work (*Sonderruhegeld*)).

In some EU15 nations when the policies for early exit and retirement were considered in conjunction, a mix of decommodification approaches was apparent in 1995. Portugal displayed a mix of labour market and rights as the guiding eligibility principles for their decommodification policies with the 'Anticipated Retirement' (*Reforma Antecipada*) policy which allowed those over the age of 55 in arduous work to retire early (Committee of the Regions, 2003), various schemes for certain occupations and an option for unemployed individuals.

Denmark and Finland's decommodification policies embodied a rights-element in terms of early exit through unemployment benefits and job search exemptions (the '50-59 rule' and the 'unemployment tunnel' (*Työttömyyseläkeputki*) respectively), whilst largely the early retirement schemes had a contribution element. In Denmark, the voluntary early retirement pension/VERP (*efterløn*) and transitional income-related pension (*overgangsydelse*) required a certain number of contribution years (although the social pension/anticipatory pension (*førtidspension*) did not) and in Finland the unemployment pension (*Työttömyyseläke*), individual early pension (*Yksilöllinen varhaiseläke*) and Early Old Age Pension (*Varhennettua Vanhuuseläkettä*) also called for long insurance records. Conversely, the Netherlands' early retirement policy was open to all with a reduced working capacity (the occupational disability programme (*Wet op de Arbeidsongeschiktheidsverzekering*, WAO)), but the unemployment benefit duration was contingent on age and contribution records. Germany's early exit schemes presented a mix of desert and rights as whilst the early retirement schemes focused on the former by requiring a certain level of insurance years (i.e. the options for women, unemployed individuals, for those with 35 years of contributions, for those with a reduced working capacity; in terms of early exit, there was the '59er' provision).

Eligibility criteria for decommodification opportunities in Belgium, France, Greece, Luxembourg and Spain represented a mix of labour market and desert. Belgium allowed individuals to exit from the labour market if made redundant with a replacement criteria for employers or from industries in economic difficulty (unemployment insurance for older unemployed individuals (*complément d'ancienneté*) and the conventional bridging pension for cases of dismissal (*prépension conventionnelle*)). There was also a flexible pension system in place in 1995 with a contribution requirement, yet in principle the ages were 65 for men and 60 for women; exit earlier than these ages would have resulted in a reduction to the final pension amount (*pension de retraite/rustpensioen*). In France the ASFNE and ARPE schemes required a certain level of contributions but also acted to reduce youth unemployment rates either through a replacement criterion or for use by industries in economic difficulty. Greece had in place two sets of rules, depending on whether individuals entered the insurance system pre- or post-1992. In terms of the system for those with contributions prior to 1992, early retirement could be entered into in cases of employment in certain industries, arduous or unhealthy work with a full pension, or in instances of long insurance durations (there was also a universal scheme but this had a reduction of 6% per year prior to the state pension age). Similarly for those entering the insurance system after 1992, early retirement also depended on employment history either in terms of contribution records or occupation. Luxembourg and Spain too focused on certain occupations and/or contribution requirements with their early exit and retirement policies.

Finally, Italy presented a real mix of eligibility criteria for early retirement and exit. Italy provided early exit universally to those over 49 through the extended unemployment benefits (the CIGs) yet there was also an additional extension under the 'Long Mobility' scheme, the early retirement Seniority Pension and '*prepensionamento*', all of which contained a contribution requirement (the latter also included a labour market element in that it was only available for certain industries). There was also a number of retirement opportunities for those made redundant from industries in economic difficulty. In addition, certain cohorts could access a 'flexible' option of exiting between the ages of 57–65 but deferring receipt was encouraged through accrual rates.

Thus the contexts in terms of early exit and early retirement varied amongst EU15 countries and therefore national active ageing approaches had very different starting points. The data demonstrate that the range of decommodification options varied, as did the eligibility criteria and their coverage. As a result, the literature explored in Chapter 2 which suggests that decommodification policies have been replaced by a focus on the recommodification of labour needs to be refined. This section will explore in more depth the second area of variance with reference to the recommodification of labour argument: the character of the reform of early exit and retirement policies, which offered the opportunity for individuals to be decommodified, and how this follows on from the national policy contexts. Indeed, just as the options for decommodification differed, so too did the character of reform and retrenchment (where applicable) as well as the subsequent recommodification policies. In short, nations' initial decommodification policy approaches (early exit and retirement) created paths along which they moved towards the recommodification of labour (the retrenchment of early exit and retirement accompanied by policies to promote employment in older age). Though policy changes were evident across EU15 nations in line with the argument that welfare arrangements now contain a work-focused element, as Pierson (1996, 2004) notes, policy contexts constrain nations' subsequent reforms to particular paths.

Generally those nations with an element of desert in the eligibility criteria for their early exit and retirement policies were bound within set reform and retrenchment trajectories. These nations either opted to remove any rights-based early exit and retirement options, raise the age and contribution thresholds for the desert-focused policies or imposed penalties on the pension incomes of those exiting; what they tended to avoid was the closure of the desert-focused routes entirely. In these instances, the retrenchment of decommodification where it has been 'earned' was perhaps not politically viable; to do so would have been to renege on existing contracts between the state and deserving individuals. Yet at the same time, where there was a mix of desert- or rights- and labour market-focused decommodification options, the latter tended to be unreformed, aside from the imposition of additional contribution requirements in the case of Greece. Instead, retrenchment in these countries took the form of shifting the rights-element towards deservingness as a criterion for exit (e.g. Portugal – see below), or retrenching those desert-focused options with the increase of contribution and/or age requirements (e.g. Belgium – see below).

Italy and Finland however present exceptions to the rule that nations retrenched rights-focused options first whilst shifting the focus to those 'deserving' of early retirement through their contribution records. In Italy, the contribution-based Seniority Pension was instead retrenched through increased contribution requirements and age thresholds yet the CIGs scheme continued to offer an extension for all workers made redundant over the age of 49. However, the system for those insured after 1995 which made exit possible between the ages of 57 and 65 represented a move from a defined benefit scheme to a 'notional' defined contribution system which meant pension payment would be directly linked to the contributions made over an individual's lifetime, thus disincentivising early exit and retirement without actively closing off these options. In addition, the Long Mobility and *'prepensionamento'* schemes both contained contribution requirements and remained unreformed.

Finland also countered the trend by not only increasing the focus on contributions but also introducing a flexible option which included a rights-based model of eligibility by 2010. Initially Finland had in place rights-based early exit through an unemployment benefit extension and desert-focused early retirement schemes. The former was retrenched through the increase of the lower age threshold of the unemployment tunnel whilst the latter were either abolished or restricted (Individual Early Retirement (*Yksilöllinen varhaiseläke*) was closed to new entrants, the Unemployment Pension (*Työttömyyseläke*) was being phased out and the age threshold for the Early Old Age Pension (*Varhennettua Vanhuuselä-kettä*) was raised to 62). However, the state pension age was replaced in 2005 by a rights-based system where individuals could retire between the ages of 63 and 68, with accrual rates that encouraged longer working lives. Yet though there were no contribution thresholds, the final pension income was contingent on contributions, with steeper accrual rates for older individuals. Thus, as with Italy, though decommodification was available on a rights-basis, individuals with limited contribution records would have received a low pension income if they opted to exit from the labour market early, thereby constraining their level of choice. This system therefore potentially transmits the disadvantage individuals face in the labour market into old age if they opt to exit, as opposed to offering choice over labour market participation.

Of those nations which strengthened their commitment to contri-bution-focused decommodification, Austria centred almost entirely on desert with regard to their decommodification options in 1995 and so the retrenchment undertaken was in the form of increased age thresh-olds and penalties for early retirement by 2010 (including the Special Regulation for Very Long Insurance scheme (*vorzeitige Alterspension bei langer Versicherungsdauer*), the unemployment pension (*vorzeitige Alters-pension bei Arbeitslosigkeit*), the early pension for those with a reduced working capacity (*Vorzeitige Alterspension aufgrund geminderter Arbeits-fähigkeit*) and the pre-retirement benefit (*allegmeine Sonderunterstüt-zung*)). In this year, Austria opted to create a new system for younger cohorts – the 'retirement corridor' (*Korridorpension*) – which though open to all of a certain age, did not represent a rights-focus in the true sense, as with Finland and Italy. Due to Austria's contribution-based pension accrual system, though individuals with disjointed work histories would have been able to exit flexibly below the state pen-sion age, the reductions incurred when combined with their limited accrual over their working lives may have made this option financially impractical.

Germany too had a strong desert-focus for its early retirement policies, though early exit was available more universally under the '59er'
unemployment provision (however, additional unemployment benefit
durations had a contribution requirement). Germany also closed off
early retirement options for younger age cohorts, after increasing the
age thresholds for early exit and retirement and imposing penalties
on those leaving the labour market prior to the state pension age. The
options available to the younger cohort in terms of decommodification
were far less extensive than for more senior individuals; the dual system
also meant that the early retirement options available to older cohorts
had a *de facto* closure date of 2016. Perhaps in Austria and Germany
retrenchment of desert-focused routes had reached their limits in terms
of what was politically viable (as Pierson (2004) highlights in his arguments regarding 'blame avoidance' and policy reform), and thus a new,
narrower system for younger age cohorts was created.

With regard to other nations with a mix of desert- and rights-focused
eligibility criteria for early exit and retirement policies, the latter
underwent retrenchment whilst the former was strengthened as the
defining principle. For example, though Denmark still embodied a mix
of desert- and rights-focused decommodification by 2010, this was a
retrenched mix of policies compared to what had been available in
1995; indeed, the rights-element was limited to an unemployment
benefit extension. Reforms in this nation had included increased age
thresholds for the rights-based early exit routes (the '50-59 rule'
became the '55-59 rule', then the '58-59' rule) and the removal of the
early retirement scheme available to those with a diminished working
capacity resulting from ill-health or social reasons (the social pension/
anticipatory pension (*førtidspension*)). The Netherlands too retrenched
the rights-based element through the abolition of the universal job
search exemption, whilst retaining the contribution focus of the unemployment benefit extension. The disability-focused early retirement
scheme (WAO) was also abolished and individuals were required to be
reassessed and if possible, placed in employment (WIA).

Where there was a mix of rights- and labour market-focused decommodification, as was the case in Portugal, the shift began towards the
inclusion of deservingness as an eligibility criterion with the addition
of contribution requirements. In the case this nation, a flexible option
of retiring between 55 and 70 was added with a contribution requirement by 2000, whilst penalties were imposed upon the early retirement scheme for the unemployed, which were then increased in 2007
along with the age threshold. In addition, the unemployment benefit

extension for the 55-plus was also retrenched through the imposition of a contribution requirement and an increase to the age threshold. The labour market-focused routes remained un-retrenched throughout the entire period. This demonstrates when a rights approach was combined with a labour market-focus, the former was retrenched through the imposition of contribution requirements, thus moving towards a desert as the eligibility principle.

In the case of nations with a mix of labour market and desert schemes, such as Belgium, France, Greece, Luxembourg and Spain, these approaches were retained over the entire period. However, in Belgium, retrenchment of the desert-based decommodification schemes was undertaken through increases to contribution requirements, age thresholds and the imposition of penalties (the contribution requirement and reduction to the final pension payment were increased for the pension de *retraite/rustpensioen* and the job search exemption was restricted), whilst the labour market-focused measures were unreformed. Luxembourg did not retrench the early retirement schemes which represented a mix of desert and labour market-foci, perhaps because the contributions required were already comparatively high at 40 years (*Pension de vieillesse anticipée, Retraite anticipée, Préretraite des travailleurs postés et des travailleurs de nuit, Préretraite-ajustement, Prerétraite-Solidaririté, Indemnité de préretraite*). Spain, on the other hand, expanded one of its desert-focused schemes by 2002, opening it to all individuals with 30 years of contributions who had been unemployed for six months (previously this had been restricted to individuals who had made 15 years of insurance contributions before 1967).

France had a contribution requirement for its early exit eligibility criteria (Special National Employment Fund Allowance (*Allocation spéciale du Fonds national de l'emploi*, ASFNE)) whilst the early retirement schemes also contained a labour market focus (the combined early retirement/hiring scheme (*L'Allocation de Remplacement Pour l'Emploi*, ARPE), end of employment leave (*Congé de fin d'activitié*, CFA), CAATA)). This focus was relatively unaltered by the reforms undertaken, though the Social Security Funding Law abolished involuntary retirement in 2008. Indeed, France's early exit options expanded somewhat with the creation of the *Allocation équivalent retraite* (AER) and the ASA (*Allocation Spécifique d'Attente*) and the age threshold for the job search exemption was lowered for those with a certain level of contributions (*Dispense de Recherche d'Emploi*). In addition, the Specific Assistance Allowance (*Allocation de solidarité spécifique*, ASS) provided additional unemployment benefits for older people.

The case of Greece is different, although this nation too retained a mix of desert- and labour-market focus over the entire period. However, in 1995, though early exit through unemployment benefits was contingent on desert, penalty-free early retirement was available for those engaged in certain occupations. Over the following ten years, desert requirements were added to these labour market-focused routes, thus restricting the number of successful candidates.

In terms of the two nations with a rights-focused decommodification approach, both Ireland and Sweden curtailed these options by 2010. Ireland completely closed off opportunities for early exit by abolishing the PRETA scheme and Sweden limited early decommodification to the flexible state pension age, having removed extended benefit durations for older individuals and the relaxed conditions for the Disability Insurance scheme.

In addition, with regard to policies aimed at retaining older workers in the labour market, incentives for the deferral of state pension receipt were advocated by the EU and as with the decommodification options, there was variation in terms of the situation in 1995 and the subsequent reforms. In 1995, Austria, Finland, Italy, Luxembourg, Sweden, Germany, Portugal and the UK provided increases to individual pension incomes if work was maintained post-state pension age. France and Spain allowed individuals with insufficient pension credits for a full pension to remain in work so as to increase their pension income. Greece did not offer an increase for deferring pension receipt in the 1990s and Belgium, the Netherlands and Ireland had no incentives for deferral throughout the entire period of investigation (1995–2010).

From the 2000s onwards, some of these nations undertook reform of this policy area. In the case of both the UK and Austria, following 2005 the incentives for deferral had been increased. France increased their incentives for deferral by 2011 and opened it up to individuals who had fulfilled the insurance requirement for a full pension. Denmark introduced an incentive for deferral in 2004 when the reduction to the state pension age came into force and Italy introduced a tax-free bonus for those with insufficient credits in this year. Spain too added an additional bonus for those with a full insurance record. Greece introduced incentives in 2005, and increased them in 2007. Portugal on the other hand added a contribution element to the incentives for the deferral of pension receipt in 2007. Luxembourg altered their rules in 2005 to instead refund individuals who continued to work post-state pension age with their pension contributions at the end of the year. Finland too reduced the bonus for deferral in 2000 only to raise it again in 2005. Sweden and Germany did not reform their incentives for deferral.

The opportunity to have some flexibility and reduce working time in the years before retirement was promoted by the EU as a means of encouraging older people to remain in employment longer. In the 2006 EC report Adequate and Sustainable Pensions, they advocated "Eroding the cliff-edge between working and retirement (which has had negative impacts for generations of workers) is an important element of increasing employment rates for older workers" (European Commission, 2006: 36). However, a number of EU15 nations did not introduce any policies to allow older people to combine work with pension receipt including the UK, Portugal, the Netherlands and Greece. France and Luxembourg conversely had a number of partial pension options available. Some of the nations had policies to allow for partial pension receipt in the mid-1990s which they made more accessible. In Sweden, the policy for partial pension receipt was available to those over 60 with ten years of contributions since 45 years of age in 1995, which was abolished in 2001 only to be reopened to all older people from 2003. Belgium, Finland and Spain relaxed the eligibility criteria. However, some nations restricted partial pension options to those with certain levels of contributions including Denmark and Italy, whilst Germany retrenched their policy by raising the age threshold. Ireland only allowed those earning less than €38 a week to combine work with pension receipt. Austria acted against the EU's active ageing approach by closing its options for part-time pensions by 2010.

In addition, with regard to policies for retaining older people in the labour market, there were reforms undertaken of state pension ages; yet in EU15 nations this presented an area of divergence over the entire period. Austria and the UK had different state pension ages for men and women throughout the fifteen years. Belgium initially had a flexible pension system in place in 1995, yet in principle the ages were 65 for men and 60 for women; exit earlier than these ages would have resulted in a reduction to the final pension amount. This was disbanded by 2000 and the pension age was gradually increasing for women to be harmonised with men. The state pension ages in Portugal were fully harmonised in 1999 and a flexible pension option allowed individuals to exit between the ages of 55–70 which incentivised longer working lives, with a minimum of 30 contribution years. For each year short of the reference pension age of 65, an individual would reduce their pension amount by 4.5%, which would be reduced for those with more than 30 years of contributions.

The Greek and Italian systems had two state pension systems in place for different birth cohorts. In terms of the Greek system, the older cohort was subject to a system with differential retirement and early retirement ages for men and women; the younger cohort's pension ages were harmonised. The Italian system increased the state pension age for both genders yet the ages were not harmonised. In addition, there was a flexible pension age instituted for individuals with less than 18 years of insurance by 1996 (between 57 and 65 years).

Luxembourg, Germany, the Netherlands and Spain retained their pension ages at 65 for both genders for the entire period (although in the case of Germany, the early retirement scheme specifically for women could also provide a *de facto* lower state pension age for those with 15 years of contributions). Ireland had two pensions in place at 65 and 66 years of age and France had a comparatively low state pension age at 60 years of age for the entire period. Denmark originally had a relatively high pension age, set at 67 in 1995 which was lowered by 2005 to 65 years.

In addition to the reform of policies to maintain older workers' connection with the labour market, the EU also advocated the re-integration of those who had exited either through unemployment or early retirement. This area of policy focus will be discussed below.

Re-integration: Convergence and divergence in policy contexts and reform

As highlighted in Section 3.4, the retrenchment of decommodification options represented the *de facto* recommodification of labour in that individuals will cease to have any other recourse but to engage in the labour market. At the same time, the introduction of ALMPs represents the *de jure* recommodification of labour. Just as the reform and retrenchment of decommodification policies varied, so too did the *de jure* recommodification policies in the form of ALMPs. The data show variation again in the context (with some nations having ALMPs for older individuals in place in 1995) and character of reform (in terms of both the timing of introduction and the type of policies implemented). This section will explore the type of ALMP introduced in terms of Dean's (2007) aforementioned taxonomy and Bonoli's (2010) classification as combined in Table 2.3. In sum, policies which adopt a work-first approach with sanctions and in-work benefits are part of the *coercive/incentive reinforcement* category. The human capital group of policies include education and training measures whilst *insertion/employment assistance* policies use direct methods to encourage re-employment such as work placements, subsidies and assistance with job search. The final set of policies includes elements such as public sector job creation and as such is labelled *job creation/occupation*.

In terms of nations with ALMPs for older individuals in 1995, France had in place an *insertion/employment assistance* type scheme which penalised employers who made older workers redundant as well as providing subsidies for this age group, and continued to introduce similar policies until 2000. Germany, Greece and Luxembourg also had subsidies in place whilst Italy, Spain and the Netherlands defended the right to work, which the latter promoted further with new anti-ageism legislation between 1996–2000. In addition, Italy and Spain provided training so as to develop human capital. Between 1996 and 2000, Germany, Greece and Spain increased the subsidies for older workers and introduced measures to improve human capital through training.

Other nations did not introduce ALMPs policies that focused on older individuals until after the mid-1990s, including Austria and Finland which created ALMPs with *coercive/incentive reinforcement, human capital* and *insertion/employment assistance* elements with Finland also emphasising the right to work through anti-ageism legislation. Belgium from the mid-1990s onwards focused on *insertion/employment* assistance and introduced subsidies and reductions for employers' social security contributions for older workers. Denmark introduced a *coercive/incentive reinforcement* element, emphasising the individual's right and duty to participate in ALMPs, in addition to improving the labour market prospects of older individuals through job integration and subsidies. Sweden too created training schemes as well as providing public sector employment, though the latter was more limited after 2000. Ireland focused on *human capital* development and *insertion/employment assistance* ALMPs through anti-ageism legislation. Portugal's approach from 1996 was limited to *human capital development* and the UK provided limited *insertion/employment assistance* measures, including an age awareness campaign and in-work subsidies for those below a certain income threshold. In addition, individuals on certain unemployment benefits could have received additional job coaching.

Between 2001 and 2010, further recommodifying policies were introduced in the EU15 nations. Austria, France and Greece provided subsidies, *human capital development* and *insertion/employment assistance* measures. However, France did abolish the *Delalande* contribution by 2010 which penalised employers who made older workers redundant. Belgium and Ireland created new subsidies and introduced anti-ageism legislation. Denmark increased the *coercive/incentive reinforcement* element of its ALMPs in addition to providing protection in employment through anti-ageism legislation. The Netherlands and Spain also encouraged individuals to engage in job search whilst providing *human capital development* measures. In addition, Spain created temporary employment placements for older individuals. Finland, Germany and Italy's ALMPs in this period focused on *human capital development* and *insertion/employment assistance* measures. The latter two nations also created job placements for older individuals. Luxembourg and Portugal's ALMP introduction in this period was comparatively limited to the creation of subsidies and human capital measures respectively. Sweden and the UK also did not undertake significant reform in this area during this period.

To return to Dean (2007) and Bonoli's (2010) typologies of welfare-to-work programmes, it becomes clear from the data regarding policies for older individuals that nations adopted a variety of approaches and thus do not neatly fit within one section of the taxonomy. Indeed, to place nations' policy approaches within one category would depict policies for older workers as overly static. Dean notes that the 'hybrid' nature of policies reflects "contested discourses of responsibility and the inherent instability of the ethical foundations of welfare" (Dean, 2007: 584). The data highlights that the policy situation was constantly evolving and shifting with the introduction of new policies and the retrenchment of existing schemes.

Convergence, divergence and overall progress towards active ageing

As would be expected, the diverse policy contexts and the divergent character of reforms and retrenchment have produced different final policy mixes in EU15 nations. Nonetheless it is clear from the data that nations have made changes in line with the EU's dual goals of retaining and reintegrating older individuals in the labour market and retrenched schemes that counter these aims, but there was still variation in the policies for work and retirement in 2010. However, policies are never static and will continue to develop and change, perhaps then converging more universally on the EU model in the coming years.

Policies for employment and retirement need to be considered in conjunction in order to address nations' progress towards the EU's active ageing agenda as this approach in itself includes both the retention and reintegration of older individuals into the labour market. In 2004 the EU categorised nations in terms of the distance from the 2001 Stockholm target, diving them into countries which were close to or surpassed 50% employment for 55–64 year olds, those which were "particularly worrisome" (Council of the European Union, 2004: 8) with less than 35% and finally those that were 'in-between'. Chapter 3 updated this categorisation with data from 2010 – the deadline for the

Stockholm target – to consider not only EU15 nations' position in terms of the 50% goal, but also the distance travelled since it was established in 2001. A number of nations were very close to or had already surpassed the 50% target in 2001 and made further progress over the following nine years (Sweden, Denmark, the UK, Ireland) whilst others which were below moved beyond it (the Netherlands, Finland, Germany). Even in terms of those nations which did not make the 50% employment target for 55–64 year olds, the consideration of distance nations had to travel is important as some nations had more ground to make up with very low employment rates in 2001 when the target was set; some then covered this ground quickly (Austria, Belgium and Luxembourg) whilst others did not (Greece, Spain, France and Italy).

To return to the updated classification outlined in Chapter 3, Group I included nations that were in 2001 close to or had surpassed the Stockholm target, including Sweden, Denmark, the UK, Ireland and Portugal. These nations' policy approaches did differ however, even if their proximity to the 50% employment rate was similar. The Group I nations can be divided into those where the high employment rate of 55–64 year olds can in part be attributed to either a lack of decommodification options and had a thus *de facto* active ageing approach, and those which retrenched their early exit options. With regard to the former, the UK was unique in that there were no state-provided early exit or retirement schemes available in the 1990s and 2000s, and therefore arguably this nation had the least distance to travel from decommodification towards the recommodification of older individuals' labour. However, the limited ALMPs introduced in the UK suggests that the approach represents the *de facto* recommodification of labour in that older individuals (without an alternative source of income) would have had no other recourse than to enter the labour market; this however does not represent a holistic active ageing approach or *de jure* recommodification of labour in that few policies were available to promote older individuals participation in the labour market. Ireland had a limited number of decommodification options available in 2001, though the one early exit scheme (the PRETA scheme) had a comparatively low age threshold as its entry requirement. However, by 2010, this option had been closed off and also some ALMPs were introduced, including *human capital* and *insertion/employment assistance* policies, yet in terms of retaining older workers post-state pension age, no incentives for the deferral of state pension receipt were created.

Sweden, Denmark and Portugal all had more early exit and retirement routes available prior to the 2000s than the UK and Ireland, which they then retrenched by 2010 and offered more opportunities for flexibility in terms of the timing and transition into retirement, including flexible pension ages, options for partial pension receipt and incentives for deferring state pension receipt. Sweden displayed a comprehensive mix of active ageing policies, including ALMPs, incentives for the deferral of pension receipt, the retrenchment of early exit and retirement and flexible options including partial pensions. However as aforementioned, the initial national context needs to be underscored. Unlike nations such as the Netherlands, Germany and Finland which were significantly below the Stockholm target in 2001 and made substantial progress to move beyond it, Sweden did not have as wide an array of early exit and retirement schemes: the only options were an extended benefit duration, disability insurance or retirement prior to the state pension age with a reduced pension amount. Thus Sweden had less distance to travel in terms of shifting from decommodification towards recommodification (but more than the UK and Ireland). In addition, there was already a bonus for the deferral of state pension receipt in place in 1995. Sweden did adopt ALMPs for older individuals over the ten year period including *job creation/occupation* schemes before 2000 and *human capital* development measures (Dean, 2006, 2007; Bonoli, 2010) as well as abolished closing the options for decommodification including early exit through unemployment benefits and disability insurance.

Denmark also had a range of decommodification policies in place in the mid-1990s, yet retrenchment was undertaken by 2010. By this year, early exit through extended benefit durations and the job search exemption had been retrenched and though in some cases individuals could retire early if they had made a certain number of contributions, they would also incur a penalty reduction to their final pension amount. In terms of ALMPs, *coercive/incentive reinforcement* and *job creation/ occupation* activation measures were introduced yet the bonuses for deferral were effectively reduced over the ten year period.

The Portuguese early retirement system comprised of four main routes by 2010: two in instances of unemployment (one via extended unemployment benefits at 57 with a contribution threshold and one at 62 with a reduction of 6% per year prior to 65), one for those employed in arduous labour and a flexible option for those with long contribution records. The policy changes undertaken are noteworthy in that over the five-year period prior to 2001 when the Stockholm target was created, Portugal added the latter flexible route. Thus as other EU15 nations were reducing their early retirement schemes, Portugal created a new, desert-focused option and applied a contribution requirement to the unemployment scheme for those over 55. Though therefore their decommodification options were retained, the shift of eligibility towards the inclusion of contribution requirements narrowed them as an option for retirement, as demonstrated by the model biography data. The onus of ALMPs available in this year focused on the development of human capital whilst deferral of pension receipt would have increased the pension by a comparatively large amount per year, but again, this was partly contingent on the individual's contribution record.

The second group of nations included in this book's fourfold categorisation were those which have since 2001 moved beyond the Stockholm target, including the Netherlands, Finland and Germany. These nations all had extensive early exit and retirement policies available in the 1990s which reflected the use of these schemes to reduce youth unemployment at that time. From the latter part of that decade and the early 2000s onwards, this policy approach shifted towards the retrenchment of decommodification options. Germany created two systems of early retirement for different birth cohorts by 2010. For those born before 1952, exit with certain levels of contributions was permitted for women, in instances of long-term unemployment or insurance durations, and where an individual's working capacity was reduced. All of these routes contained contribution requirements and reductions to the final pension amount. For younger age cohorts, the scheme for women was closed and the age thresholds for the disability and long insurance routes were progressively raised. There was a job search exemption in place for those over the age of 58.5, but this was being phased out as of 2006. In terms of the ALMPs, Germany had focused on *job creation/occupation* and *human capital* policies and anti-ageism legislation, whilst individuals would have been able to increase their final pension amount if they continued to work beyond the state pension age.

Though Finland had undertaken a great deal of retrenchment with respect to its early retirement schemes in place by 2010 (and perhaps accordingly had surpassed the Stockholm target) the unemployment tunnel was retained, providing early exit from the age of 57. From this, individuals could have moved onto the contribution-based unemployment pension, which contained a penalty of 4% per annum. In addition, there was an early pension for farmers and a general scheme with a reduction of 6% per annum. The case of Finland demonstrates the importance of context and the mapping of reforms; if assessed solely in terms of the early exit and retirement routes in 2010, it would appear to have a long way to go before it embodied the EU's active ageing agenda. However, compared to the extensive list of routes from the labour market for older individuals in 1995, those available in 2010 represent a significant reduction and largely had also been retrenched through increased age thresholds. In addition, Finland also provided a bonus for the deferral of state pension receipt and ALMPs including human capital development, subsidies and anti-ageism legislation. Therefore though Finland still had a comparatively large amount of decommodification options available in 2010, these were greatly reduced compared to the situation in the 1990s and the pension system had been reformed to encourage longer working lives through age-related accrual rates.

The Netherlands could also be classed with Sweden and to a degree Denmark as a nation with less ground to cover towards the development of active ageing policies in terms of the absence of early exit and retirement options. However, unlike Sweden and Denmark, the Netherlands was below the Stockholm target for older peoples' labour market participation and only just above the EU average of 38.8% in 2001 (Eurostat, 2011). In the case of the Netherlands however, this was perhaps because private occupational schemes (VUT) allowing early exit were extremely popular. Yet of the state-provided decommodification options available in 1995, by 2010 for those over 57.5 years of age, there only remained the exemption from actively seeking work and an unemployment benefit extension for those with a certain level of contributions; the disability route (WAO) had been replaced by the work-focused WIA. The ALMPs introduced included anti-ageism legislation, human capital development and training, yet no bonuses for deferment were available.

With regard to the EU15 nations which had by 2010 not reached the Stockholm target, two sets can be identified: those which had made significant progress when their low employment rates in 2001 are considered, and those which made more limited progress (below the average for all EU15 nations (+9.6%)). This thus takes into account not only national progress in relation to the 50% target, but also the distance travelled as it became apparent that there were both marked differences between those who made significant gains over the nine years and those whose progress remained relatively limited.

Group III nations include Belgium, Austria and Luxembourg who all raised the employment rate of 55–64 year olds beyond the EU15 average increase (+9.6%) but still remained below the Stockholm target of 50%. In terms of these nations which had yet to reach the Stockholm target but had made significant increases in the employment rates for older workers, when their starting point in 2001 is taken into account, Austria retrenched the decommodification policies available, introduced a range of ALMPs and provided incentives for deferral. With regard to exit, a dual system for different age cohorts had been created in accordance with which those over 50 in 2005 would have been able to retire early if the individual had been employed in heavy industry or had a long insurance duration (though the latter was being phased out); for those under 50 in 2005, early exit from the labour market would have incurred both a loss of pension income in terms of contribution years and also a penalty deduction. In addition, there were incentives for the deferral of state pension receipt and ALMPs encompassing Dean (2006, 2007) and Bonoli's (2010) *insertion/employment assistance, human capital development* and (pseudo) active *job creation/occupation* principles.

Some nations in this third group adopted a more 'mixed' active ageing approach, denoting the retention of certain decommodification options whilst at the same time retrenching others and introducing ALMPs. These nations made significant moves towards the Stockholm target from relatively modest employment rates for older workers in 2001. Belgium had in place early exit through extended benefit durations for those with extensive contribution records, as well as early retirement in instances of dismissal and for long insurance records. Individuals who had become unemployed after July 2002 had to engage in job search if under the age of 57, and ALMPs were in place that encompassed subsidies, job placements and anti-ageism legislation. However Belgium had not implemented incentives for the deferral of state pension receipt.

Luxembourg too had not managed to exceed the 50% employment rate target for older workers by 2010, perhaps due to the range of early retirement policies – totalling seven schemes – which had not been retrenched over the ten year period. Yet in comparison to other nations, the desert-focused options had relatively high contribution requirements. However, Luxembourg appeared to counter the EU's recommended active ageing approach by abolishing the bonus for deferment by 2010. The ALMPs available in this year included subsides and were also limited in comparison to other EU15 nations.

There was a final group of nations whose progress towards the Stockholm target was more modest, neither attaining 50% nor the average increase for EU15 nations (+9.6%). These nations tended to offer more in the way of alternatives to labour market participation, although retrenchment was undertaken, and the options for exit still tended to be utilised as a means of managing the composition of the labour force. France retained a range of labour market-focused decommodification opportunities, though by 2010 ALMPs had been introduced. France had a fairly extensive array of early exit and retirement schemes still in place in 2010.

Individuals over the age of 57 were exempt from looking for work as well as those who had made 160 insurance quarters at the age of 55. There was also a supplement to the unemployment benefit for those over a certain age and level of contributions. In terms of early retirement, there were two partial schemes, two routes for those employed in arduous labour and a policy for firms facing economic difficulties to shed their older workers. With regard to the ALMPs available, France focused on anti-ageism measures and subsidies. In addition, in the case of France the deferral of state pensions was limited to individuals who had yet to secure the full pension amount through their contributions.

Greece too retained the labour market-focused early retirement options, though contribution requirements had been applied to all routes, some of which contained penalties. Greece also had different systems of early retirement for different age cohorts. For those who entered the insurance system after 1993, early retirement was available with either reductions or without, for certain occupations and contribution records (four routes available in total). Individuals who entered the insurance system pre-1993 could similarly access both reduced and unreduced early pension, though there were more schemes, 11 in total. These routes all contained a contribution requirement by 2010, thus limiting older individuals' access. In addition, a bonus for deferment had been introduced, as well as ALMPs including job creation, human capital and anti-ageism legislation.

Spain too had failed to reach the Stockholm target and had not made the same level of progress as other EU15 nations. In Spain in the mid-1990s, there were three main early retirement routes, two of which were for instances of industrial restructuring and one for long insurance durations with penalties. In addition, there were also schemes limited to particular industries and the '*Relevo*' (take-over) scheme (Forteza and García-Zarco, 1998) which included the requirement that the employer replace the older person with someone from the unemployment register. In terms of reform, these routes were retained but the scheme for long insurance durations was made more flexible with high penalties and contribution requirements. With regard to early exit through unemployment benefits, this option was retrenched through the requirement to actively seek employment. ALMPs introduced by 2010 included anti-ageism policies, human capital development measures and job placements. However, the bonus for deferring state pension receipt was low compared to other nations at only 2% per year.

In 2010 the Italian system offered early retirement for long insurance durations and to individuals who had been made redundant. With regard to the state pension system, three options were in place, one for those who had made 18 years of contributions prior to 1995 were subject to the rules of the *Amato* reform; one for those with less than 18 years of contributions would partially receive their pension in accordance with the old and new systems in proportion to the years of contributions pre- and post-1995; and finally, those entering the system after 1995 would be subject to the new rules of the *Dini* system which would be fully in place by 2030–5. The state pension ages for the former had been increased to 60 for women and 65 for men whilst the newer system offered exit between the ages of 57 and 65, with the accrual rates to encourage longer working lives. Individuals deferring pension receipt would have received a tax-free subsidy, equivalent to their social security contributions. The ALMPs available included anti-ageism measures, job placements and human capital development through training.

Thus when considering policies for older individuals in conjunction, the picture is very complex; nations may retrench or retain early exit and/or retirement schemes on the one hand whilst introducing ALMPs and flexible retirement options. To return to the literature explored in the first chapter, the notion that nations are shifting from decommodifying welfare arrangements towards the recommodi-fication of labour needs to be more nuanced and refined to take into account the very different policy starting points in terms of the former which create challenges in path of reforms towards the active ageing agenda. The recommodification of labour argument does not explore the nature of the reform and retrenchment of decommodification policies; indeed, it underplays the resilience of many decommodification options, and that often they co-exist with newer, recommodifying policies. The examination of the initial decommodification approaches in the EU15 nations demonstrated that reform and retrenchment towards the *de facto* recommodification of labour was bound by these original contexts, thus reflecting the arguments of Pierson (1996, 2004). Yet to stress the case of Pierson too strongly may also under-play the degree of change EU15 nations underwent, and the progress made towards the EU's active ageing agenda. The empirical data demonstrates EU15 nations have all made changes to their policies for older individuals in order to encourage them to enter and remain in the labour market, yet these changes have not resulted in linear convergence. The diver-gent reforms and final policy pictures in 2010 reflect the differ-ent policy contexts prior to the establishment of the EU's active ageing agenda in the early 2000s, which dictated the distance nations had to travel towards the EU's active ageing goals, and the shape that reform and retrenchment could take.

The recommodification of labour literature can be further refined when applied to the data regarding the policy treatment of older individuals. The depiction of a broad shift from the decommodification to the recommodification of labour through social policy obscures a great deal of divergence within both elements. The availability and eligibility criteria for the decommodification of labour varied significantly in EU15 nations throughout the period, as did the subsequent policies available which focused on recommodification. The next section will demonstrate how the reforms and retrenchment undertaken had different effects at the micro-level, which is an important facet to be added to the recommodification of labour arguments: which groups are recommodified whilst others can be decommodified?

8.2   Policy convergence and divergence: The micro-level

As demonstrated in the previous section, EU15 nations' policies did not move towards an identical active ageing approach, in part due to their different policy legacies. What is also apparent is that at the micro-level, active ageing policies within these nations did not represent a homogenous experience for older people. The data highlight the need for more refined versions of the recommodification and reserve army of labour arguments. Not *all* older individuals were equally subject to the recommodification of labour, or were being drawn into the labour market in a period of low unemployment; indeed for some the opportunity for decommodification was limited or absent from the outset. From the period prior to the establishment of the Stockholm target until 2010, some sub-groups within the category of 'older age' had more choice with regard to their labour market participation pre-state pension age in that decommodification options remained available whilst ALMPs for this age group were introduced. At the same time, certain groups' abilities to be decommodified were being retrenched whilst ALMPs focused on the recommodification of labour. Therefore the arguments that individuals are being recommodified or that all older people are part of the reserve army of labour, drawn into and expelled from the labour market as the economy requires obscures a great deal of variation: not all individuals had the option of decommodification, and some retained this ability and were less subject to the logic of recommodification. Thus to paraphrase Orwell (2008: 133), in EU discourse all individuals should age actively, yet when examined at the micro-level, some are required to age more actively than others. This differential policy treatment of sub-groups within the older age cohort will be explored in the following section in terms of the impact of labour market participation, age and gender on policy choice.

Labour market history

The data utilising the model biographies demonstrate the importance of labour market opportunities in shaping the experience of old age, in line with the political economy of ageing arguments. Whilst it could be argued that to provide increased choice *vis-à-vis* labour market participation for those with extensive contributions is not necessarily inequitable – after all, those who have worked for the longest periods of time may deserve to be decommodified at a younger age – this assumes that opportunities to participate in the labour market are bestowed equally upon all citizens. Anderson (2004) and White's (2000, 2004) arguments regarding workfare can be applied to the notion of 'deserved' decommodification: states that demand a certain level of contributions before an individual 'deserves' decommodification need first to ensure equality of opportunity for individuals to secure these contributions. With regard to labour market participation, gender is particularly relevant in some nations due to the unharmonised age thresholds and the impact care has on employment prospects. Leaving gender aside until the section below, labour market history is not only a pertinent area for women; other social characteristics also impact upon the ability to participate in paid employment, such as disability, recurrent health problems, discrimination on the grounds of age or race, as well as the devaluation of skills and the decline of certain industries.

Long, uninterrupted careers were an important determinant of decommodification opportunities in some nations prior to the establishment of the Stockholm target in 2001, and became increasingly significant by 2010. To return to the schema outlined in Section 3.6, as the data comparing *Laurent* and *Sasha* (55 and 63 respectively with 35 years of employment contributions) with *Jean* and *Jude* (again, 55 and 63 respectively but with disjointed work histories) demonstrate, the division of deserving and undeserving was often made in accordance with contribution records. This was particularly the case in Austria, where all decommodification options contained a threshold of insurance contributions with the exception of the job search exemption and in instances where disability was the result of a work accident. This desert-focus was increased during the 2000s with the abolition of the job search exemption and early retirement scheme for those with a reduced work capacity. In addition, the desert-focused early retirement policies were also retrenched through increases to the age thresholds and the imposition of penalties. In other nations, such as Denmark, Portugal and Greece, the focus on desert was introduced or increased over the ten-year period, thus barring *Jean* and *Jude* from many decommodification options. At the same time, ALMPs offered increased possibilities for recommodification.

However, labour market participation not only affected an individual's ability to exit early; it also impacted upon the level of pension income in many nations, which could have constrained the decision to remain in the labour market. In nations where the final pension income was contingent on contribution years, this provided an additional consideration with regard to *Jean* and *Jude's* exit. Austria, Belgium, France, Germany, Italy, Ireland, Finland, Greece, Ireland, Luxembourg, Portugal, Spain and the UK all had first tier[2] pension systems linked in some way to contributions or earnings with only low-level social assistance pensions for those with no contributions, whilst Denmark, the Netherlands and Sweden[3] and all had a first-tier, basic flat-rate citizenship pension with additional supplementary schemes. This adds an additional consideration for older people considering exiting the labour market: if their pension amount is contingent on contribution records, individuals with limited labour market opportunities may receive significantly lower pension incomes than others within their age cohort with fuller employment histories. In nations where deferring pension receipt was not possible (Luxembourg, the Netherlands and Ireland), the opportunity to try and increase their final pension amount by retiring later would also not have been possible.

In terms of occupational class and its relationship with employment, in some nations early exit and retirement were occupationally segmented. First, the notion of a 'full' career varied in accordance with different industries, perhaps because of negotiations between the state and the social partners (Ebbinghaus, 2006; Immergut et al., 2007) or because they represent the opportunity for companies to shed older workers (Jacobs and Rein, 1994). The labour-market elements of decommodification proved resilient in Belgium, France, Italy, Portugal, Spain, Luxembourg and Greece. Second, supplementary pensions, whether provided by the state, employer or private sector, are not equally distributed, either in terms of availability or in terms of the ability of the various sub-groups within the older age cohort to make sufficient contributions.

Age

Age is clearly important when addressing policies for older workers. However it became clear that policy options varied according to age *within* the older age cohort, i.e. those aged 55–64. The comparison of policies available to the model biographies aged 55 and 63 not only demonstrated the significance of labour market history in the form of contribution and insurance records, it also underscores the impacts of age and gender on decommodification opportunities. The data highlight how the policy treatment of *Laurent* and *Jean* was different from that for *Sasha* and *Jude* on the basis of age.

For example, age was particularly pertinent in France with regards to the state pension as *Sasha* and *Jude* would have been able to retire at 60 throughout the entire period. In the case of the Netherlands, early retirement had relatively low age thresholds, whilst early exit through unemployment benefits would have been limited for *Laurent* and *Jean* due to their age. In Luxembourg throughout the entire period the 50-plus model biographies' ability to be decommodified through extended benefit durations was contingent upon their contribution records as their age barred them from the mix of desert- and labour market-focused early retirement schemes available to those over the age of 60. In the case of Sweden, *Laurent* and *Jean* were too young to retire early, but could access extended benefit durations and disability insurance in 1995 whilst *Sasha* and *Jude* could retire at their discretion via the flexible option or the Disability Insurance scheme. Age became more significant in Sweden in that the age threshold for the unemployment benefit extension was raised by 2000 to 57 before being abolished altogether by 2010 and the more relaxed conditions for the disability insurance for those over 60 were withdrawn in 1997. With regard to Germany, though *Laurent* and *Jean* were too young to retire or exit early, their unemployment benefits were contingent upon their age and contribution records. In this nation, retrenchment took the form of increased age thresholds and the imposition of penalties for the early retirement routes, though *Sasha* and *Jude* were unaffected by this change as they were older than the new ceilings.

In addition, retrenchment in many nations often took the form of increased age thresholds, thereby curtailing the model biographies' decommodification options, as was the case with Sweden. In those nations where decommodification opportunities had a mix of labour market- and desert-focused decommodification policies, such as Belgium, reform consisted of retrenchment of the latter routes through the increase of age thresholds, imposition of penalties or total closure of routes, whilst the former were largely retained. As a result, all model bio-graphies would have been able to exit or retire early if employed in certain occupations or industries in difficulty. Similarly, Italy reformed its desert-focused scheme (Seniority Pension) by gradually increasing the age and contribution threshold whilst retaining its scheme for those made redundant, though this too required 15 years of insurance contributions. Denmark had in place relatively low age thresholds for decommodification in the 1990s, yet within ten years, these limits had been raised to bar *Laurent* and *Jean*. Finland too initially provided *Laurent* and *Jean* the option of remaining on unemployment benefits whilst for *Sasha* and *Jude* the early retirement routes exemplified a mix of rights- and desert as their eligibility principles. By 2005, the age threshold for the rights-based decommodification option for *Laurent* and *Jean* had been closed off whilst recommodifying ALMPs had been introduced. In the case of Austria, age and gender combined to allow retirement and exit for *Laurent* due to their contribution record in the mid-1990s. As with the Finland and Denmark, retrenchment of these options took the form of increased age thresholds, whilst the onus on desert was increased for *Sasha* and *Jude*.

Age was also important with respect to cohort membership in Italy due to the *Amato* and *Dini* reforms as people were subject to different pension rules depending on their level of contributions at the time of the reform; a year's difference in birth could result in a variance of 15% replacement rate for the final pension amount. For example, one person born in 1957 and another born in 1958 with the same level of contributions would respectively be subject to the rules of the *Amato* reform and a mix of the *Amato* and *Dini* systems and as a result, they would also receive their pensions at 80% and 65% of the replacement rate correspondingly (Billari and Galasso, 2008). Greece, Austria and Germany also had in place early exit and retirement schemes for differ-ent birth cohorts, with those available to younger people more restric-tive. This allowed the latter three nations to reduce future access to early exit and retirement without completely removing the acquired rights of current cohorts of older people.

Gender

The model biography data highlight the importance of gender in shaping the policy options available to older people. As aforemen-tioned, many nations used contribution records as a means to calculate pension amounts or to control access to early exit and retirement. A number of factors inhibit labour market participation for women, including care provision and employers' expectation of care provision (which may limit employment opportunities) and they may find 'earning' decommodification more problematic (Knijn and Kremer, 1997; Himmelweit and Land, 2008; Crompton et al., 2003). With regard to care, Bettio and Plantenga (2004) argue nations' policies for social care influence the extent to which the family, and principally women, act as the provider. As a result, the degree to which women's work histories are interrupted by the provision of care in part is dictated by the level of state involvement in this area (Bettio and Plantenga, 2004; Millar, 1999; Pfau-Effinger, 1999). Again, without equality of opportunity for individuals to secure the contributions required for decommodification, the onus on ageing actively will fall disproportionately on certain groups; providing care can unequally bar women from securing sufficient pension credits to have the same level of choice over their labour market participation as their male counterparts. In addition, Anderson argues unpaid care is undervalued by states, which fail to acknowledge its status as "socially necessary labour" (ibid: 248). Thus pensions do little to redress the lack of recognition unpaid care receives as an area of labour during the period in the lifecycle designated for 'work'; indeed, they persist in devaluing labour within the family (Arber, 2006; Ginn and Arber, 1995; Price and Ginn, 2006; Samorodov for ILO, 1999: 32). Therefore the experience of ageing is bound up with previous interactions with the labour market and state policies and this is particularly pertinent with regard to women and decommodification.

Thus there are two ways systems which focus on contribution records whilst at the same time disproportionally placing the burden for care on women (at the same time as failing to acknowledge the important role of care provision through pension credit accrual) can have unequal effects along gender lines. First, in terms of state pensions at regular retirement age, a disjointed work history in some nations will reduce an individual's pension income if the final amount is indexed according to contributions – in nations where there are insufficient credits for periods of caregiving and state-provided alternatives to familial care, women may be at risk of lower pensions than their male counterparts. In pension systems where the final pension amount was contingent upon accrual over the individual's working life, women again may have made less contributions due to time spent providing care, dubbed the 'family gap'. When combined with the 'gender gap' (Evandrou and Falkingham, 1995) with regard to pay, this reduces women's choice regarding exit, as remaining in work is the only way they can secure an acceptable level of income, or leads to dependency on their spouses.

In terms of this first area, in Italy for example the first and primary pension pillar was based on contributions. Though there was a secondary pillar for those with insufficient contributions, this was means-tested and at a low rate (Ferrera and Jessoula, 2007). Whereas three-fifths of men in Italy had 30 years of social insurance contributions, only one-fifth of women did in 1993. Women tended to have disjointed employment records and thus 54% of women who received pensions had less than 20 years of contributions compared to 19% of men. As a result, they were reliant on the voluntary scheme for those with incomplete records of which women made up 82% of contributors. There were no credits to recognise maternity leave or care for a family member. Thus public pensions in Italy discriminated against women in that they took little account of their differential relationship with the labour market (Walker, 1993). Coupled with this, a number of authors have identified Italy as a nation where the state has a limited role in provision of care; instead the family, and principally women, provide care for younger and older family members, inhibiting their labour market participation (Bettio and Plantenga, 2004; Anttonen and Sipilä, 1996; Millar, 1999).

In addition, in many EU15 nations, pension age thresholds as well as those for early exit and retirement are not harmonised. As a result, in some nations, if *Sasha* and *Jude* had been female they would have been over the state pension age whereas the early retirement age thresholds may have allowed a female *Laurent* and *Jean* to be decommodified. However, where pension ages are unharmonised, women effectively have less time to accrue a sufficient pension income before they reach the state pension age for their gender. In terms of this intersection between gender and age, the UK had in place unharmonised state pension ages for men and women, thus allowing *Sasha* and *Jude* to retire at 60 if female. Greece too retained unharmonised ages for men and women for both the state pension and the early retirement routes, although this is not the case for cohorts who entered the system post-1993. Given that these nations had contribution requirements for the retirement pensions and women's increased likelihood of periods of caregiving, these lower age thresholds in effect provide females less time to accrue a sufficient pension income. The same conclusions could have been made about the Belgian and Portuguese state pension ages due to their systems of accrual yet the harmonisation of male and female state pension ages was being undertaken gradually.

If the first area where gender affects decommodification opportunities is around conventional state pension receipt, the second area of constraint is related to early retirement schemes. In some nations insurance contributions determine access to decommodification through early exit and retirement schemes, women may be more likely to be required to age actively through prolonged labour market participation. In nations where early exit and retirement required a certain number of contributions, lower age thresholds for women effectively meant they had less time to accrue sufficient years; as with the state pension ages, even where age thresholds for early retirement were harmonised, contribution requirements may have reduced women's opportunities for decommodification as they are more likely to have disjointed work histories due to their caring role. In addition, many of the nations with a desert-focus or element in their decommodification policies undertook retrenchment in the form of penalties for those accessing these routes in the form of percentage decreases to the final pension amounts. In these nations, penalties in addition to income lost through accrual would have been borne more easily by those with extensive contribution records. Again, this highlights the importance of labour market participation in shaping the experience of old age: decommodification can be afforded by those with long work histories, who are less likely to be female. In addition, individuals with recurrent health problems are also disadvantaged in terms of these desert-focused routes, and the penalties they would impose.

For example, in the case of Austria, the unharmonised age thresholds for the early retirement schemes effectively meant women had five years less than their male counterparts to meet the contribution requirement. In addition, the pension amount was linked to individuals' contribution records yet the lower age thresholds for women therefore meant they would have less time to accrue a good pension income. This when combined with women's increased probability of having a work history interrupted by care produces a low pension income when compared to their male counterparts. Though the state pension ages in Germany were harmonised, there was an early retirement route for women that contained a contribution requirement. Women opting to take this route in 1995 would only have reduced their pension amount in terms of lost insurance years between the age threshold and the state pension age; however, an additional penalty was applied to this route by 2000. Greece too had lower early retirement age thresholds for women born in particular cohorts and like Germany, the subsequent application of contribution requirements would effectively have given women less time than their male counterparts to accumulate insurance years. However, in the case of Greece, new early retirement policies were introduced by 2000 for women with dependent children, yet these too contained a contribution requirement.

Thus gender constrains female decommodification – both early and regular retirement – in two ways. First in a practical sense, women's interrupted work histories that result from caring may bar them from accessing decommodification policies with contribution thresholds. Second, women may be implicitly deterred from decommodification when final pension amounts are linked to contributions, which a limited work history may make financially untenable for women. As a result, married women may become dependent upon their partners' pensions to provide a reasonable standard of living in old age (Meyer, 1998).

The data from the model biographies highlight that policies for older individuals create divisions within this age cohort along the lines of work history, gender and age. As a result, policies for work and retirement play a role in not only perpetuating disadvantage accrued over the lifecourse, but also exacerbating these differences through the unequal policy choices presented to the sub-groups within the older age cohort. Though nations' active ageing policies aim to address intergenerational equity by ensuring younger cohorts are not burdened by passive older generations, they could contribute to intragenerational inequity between older individuals.

8.3   Summary

EU15 nations have all made some moves towards the EU's active ageing agenda, and thus the recommodification of older individuals' labour. However, these moves should not be taken as indicative of linear convergence, as nations still display a significant amount of divergence with regard to their active ageing approaches. This chapter outlined the two main levels of variation within EU15 nations' active ageing approaches: at the macro-level in terms of their policy contexts, the character of reforms, final policy mixes; and at the micro-level in terms of the options available to individuals within the older age cohort. Thus the recommodification of labour argument can be refined to include considerations of the original policy contexts, in terms of the decommodification policies available and their impact on subsequent reform trajectories, as well as the differential policy treatment of individuals.

9
Conclusion

This volume has explored the policy treatment of older individuals in EU15 nations from the mid-1990s to 2010 in order to examine the implementation of the EU's active ageing agenda in these countries. It began by providing the macro-context in terms of the recommodification of labour literature which argues decommodifying policies have been retrenched whilst the focus has shifted towards the provision of welfare through the (labour) market. However, there is also a body of literature which suggests historically older individuals have been drawn into and expelled from the labour market when required by the economy as they make up part of the 'reserve army of labour' (Phillipson, 1982, 2005; Taylor and Walker, 1996). Thus, it is argued early exit and retirement schemes in many nations were introduced to reduce youth unemployment during the economic crises of the 1970s and 1980s and the subsequent shift towards 'active ageing' through labour market participation can equally be seen as the result of predicted labour shortages and demographic change.

Though there may be disagreement between the recommodification and reserve army of labour literatures – the latter suggests the move towards activation policies is part of a cyclical process whilst the former suggests the shift is the result of endogenous and exogenous pressures on decommodifying welfare arrangements – both approaches focus on the macro-level. As a result, they do not address differential policy treatment at the micro-level in terms of the division of welfare clients into 'deserving' and 'undeserving'. The third main literature employed by this book was therefore the political economy of ageing perspective, which highlights policy's role in creating and perpetuating differences within age cohorts. Thus not all older individuals would have been decommodified or expelled from the labour market and equally not everyone's labour will be subject to recommodification.

In exploring these literatures, this volume first addressed whether EU15 nations were moving towards the EU's active ageing agenda in the period from the mid-1990s to 2010 (i.e. the phase prior to the launch of the EU's active ageing approach until the deadline for the Stockholm and Barcelona targets[1]) which can be viewed as indicative of the shift away from the decommodification of labour towards its recommodification, as well as part of a broader trend of older individuals' inclusion in the reserve army of labour. In terms of the policy areas and reforms under investigation, this book focused on those advocated by the EU, including the retrenchment of early exit and retirement schemes, the creation of ALMPs focused on older individuals, the increase of state pension ages and the introduction of incentives for deferral. The book sought not only to examine EU15 nations' policy mixes at two points in time and provide a contrast, it also strove to examine the character of reform and retrenchment and address whether certain policy legacies bound subsequent policy changes within set parameters, as implied by the new institutionalist literature (Pierson, 1996, 2004; Streek and Thelen, 2005).

In addition, influenced by the political economy of ageing literature which argues the experience of ageing is not homogenous and filtered through national policies, 'model biographies' were employed to address EU15 nations' differential policy treatment of older individuals. The recommodification and reserve army of labour literatures both focus on the macro-level; they do not address how state policy interacts differently with different groups of people. Policy has traditionally divided recipients into those 'deserving' and 'undeserving' of decommodification and this book addressed which older individuals were placed in which category, and thus were less subject to the 'active ageing' agenda. The eligibility criteria of policies for older people were examined so as to establish the policy options available to different sub-groups within the 55–64 age group, and the degree to which they altered after the establishment of the Stockholm and Barcelona targets. As noted in Section 2.5, a lack of choice with regard to the labour market, either in terms of participation or exit, cannot be viewed as positive. Employment can provide self-worth, meaning and financial security yet equally, it can be insecure, poorly paid and dehumanising. By the same virtue, exit from the labour market can provide the opportunity for self-fulfilment and for some can be financially rewarding, yet it can also be an extremely negative experience in terms of a loss of income and identity.

Thus *choice* is key in relation to the labour market, and as Leisering (2003: 19) notes, individual choice is embedded in structures and institutions and the welfare programmes they create "provide competencies, resources, opportunities, and individual rights that empower individuals outside and inside of the market and the family". The use of model biographies allowed for the exploration of the degree of choice older individuals had with regard to the labour market, and through the examination of eligibility criteria, it also addressed the way policy constructs the experience of ageing in terms of who is able to become 'passive' through decommodification, and who is in need of 'activating' through recommodification. In addition, this book provided an examination of the impact of labour market history on the experience of ageing, thereby exploring the notion that disadvantage can be carried through into old age.

The focus was therefore not on causation, in terms of the impetus behind nations' policy changes and despite acknowledging the new institutionalist literature, it was also not on the processes of change. Nor did it aim to address outcomes at either the macro- or micro-levels in terms of reduced state expenditure on this age cohort or individual financial losses and gains. Instead the emphasis was essentially on three questions: Are all nations converging towards the EU-vision of active ageing? What was the character of reforms undertaken in these nations? Finally, are all individuals within the category of 'older age' equally subject to active ageing policies?

In answering these three questions, it was first important to consider the policy contexts within which the active ageing agenda was situated. In the mid-1990s, prior to the renewed focus on ageing actively through engagement with the labour market, many EU15 nations provided early exit and retirement options. Early exit via extended unemployment benefits and job search exemptions as well as *de jure* early retirement policies allow for the 'decommodification of labour' as they permit individuals to live without recourse to the sale of their labour on the market. As aforementioned, the *de facto* early exit options such as job search exemptions and extended benefit durations need to be considered in conjunction with the *de jure* early retirement routes to establish a true sense of nations' decommodification approaches. However, in some nations when these policies are combined, they represent a site of contradiction and tension (for example in Germany, Greece and Portugal in the mid-1990s, when early exit and retirement were considered together, the focus of decommodification in terms of eligibility criteria is mixed).

The guiding principle of these policies' eligibility criteria has allowed for the classification of opportunities for decommodification. Three main policy approaches to decommodification have been identified in this book: desert-, labour market- and rights-focused eligibility criteria. Broadly speaking, for nations where decommodification was for the large part contingent upon the individual's work record, the approach is labelled 'desert-focused', with desert here defined as "a backward-looking concept" (Miller, 1989: 93), contingent on previous behaviour. When decommodification was universally available to all citizens above a certain age (prior to the state pension threshold), these nations adopted a 'rights-focused' decommodification approach. Finally, those nations where decommodification was bestowed to individuals in certain industries and occupations, contains criteria that allow firms to replace the older individual with a young person from the unemployment register or are for use in times of economic crisis, these nations had a 'labour market-focus'.

This context is important as the policy legacies in these nations created the parameters for subsequent reforms in line with the active ageing agenda, as well as the distance nations had to travel towards 'activating' older workers (i.e. those nations with many early exit and retirement policies had more ground to make up than those with few or no options for targeted decommodification options prior to the state pension age). These particular approaches to decommodification had their own particular pathways to active ageing and the nature of the early retirement routes available prior to the launch of the Stockholm target in 2001 shaped the path retrenchment or reform took over the following nine years. The picture is therefore more complex than linear convergence towards the EU's active ageing approach, representing the recommodification of labour. However, change has occurred, perhaps to a greater degree than the new institutionalist literature would envisage without the necessary 'critical juncture' (Pierson, 2004).

Retrenchment itself in the EU15 nations came in a variety of forms, including the introduction or increase of penalties for leaving the labour market before state pension age, raising age thresholds and contribution requirements as well as the reform of accrual rates to incentivise longer working lives or the total closure of early retirement schemes. This section again groups nations in terms of the guiding principle upon which decommodification was bestowed in the 1990s: whether they were desert- (categorised as policies with a contribution threshold), labour market- (as policies available to certain sectors or with replacement requirements), rights-focused (available universally to all citizens above a certain age), or indeed a mix of these. The UK is also distinct in that it did not have any state-provided early retirement schemes in place for the entire ten-year period and therefore is not included in this section. With regard to the character of early retirement reforms undertaken, nations with a desert-focused element to their decommodification opportunities undertook retrenchment by initially raising the age and contribution thresholds and the imposition of penalties for those taking the option of decommodification, whilst also limiting access to universal schemes. In the period following the Stockholm target (2001 onwards), these nations either introduced a rights-based model whereby the individual's pension age was flexible and the amount was dependent upon their accrual of insurance years for the entire population (Finland) or limited these changes to younger age cohorts (Austria). Germany too limited certain decommodification options to older age cohorts, which therefore produced *de facto* closure dates. Nations where there were several decommodification options, some of which had rights or desert in their eligibility criteria, retrenched policies which focused on the former (Germany, Denmark and Netherlands), thereby emphasising decommodification in older age had to be earned.

In terms of nations with universal schemes alone, Sweden and Ireland retrenched their rights-based decommodification options. Ireland closed off its universal early exit scheme by 2010 and Sweden limited early retirement to a flexible scheme which incentivised longer working lives through the pension credit accrual arrangements. Thus these policy reforms would imply that recommodification or 'active ageing' is for *all* older people and some are not compelled to be more active than others.

The nations with a labour market-focus tended to retain these schemes, largely unretrenched as was the case of France. In addition, these nations also added either a desert-element to their labour market-focused early retirement schemes (Greece) or introduced additional routes with contribution requirements. In terms of the latter, Portugal initially focused on the labour market and rights with its early retirement policies, but introduced a contribution requirement to the early exit option and created a new desert-focused flexible route, thus shifting this nation to include a desert-element to the early retirement policies. The rights-focused scheme for those who were unemployed was retrenched at the same time to include a greater penalty for early retirement and a high age threshold.

Aside from the *de facto* recommodification of labour through the retrenchment of early exit and retirement, the EU advocated the creation of policies for *de jure* recommodification through the introduction of active labour market policies (ALMPs) for older people. The general conclusion from the data could be that EU15 nations have indeed adopted ALMPs tailored for older individuals that could be considered part of a move towards an active ageing agenda yet once again this is too simplistic, ignoring the rate and type of policies adopted in the various nations. What the data demonstrate is that nations adopt different mixes of ALMPs for older individuals, and thus do not sit neatly within one-quarter of Dean's (2007) or Bonoli's (2010) aforementioned taxonomies of welfare-to-work approaches. In terms of timing, for the great majority 1996–2000 was the main period of activity in this respect, with some nations maintaining their efforts whilst others reduced ALMP introduction between 2001 and 2010.

Similarly, in terms of policies to encourage longer working lives, incentives for deferring pension receipt were another area of policy divergence. Some nations did not introduce any incentives over the entire period under investigation (Belgium, the Netherlands and Ireland). Of those that did, there were different starting points as in the mid-1990s, Austria, Finland, Italy, Luxembourg, Sweden, Germany, Portugal and the UK provided increases to individual pension incomes if work was maintained post-state pension age whilst France and Spain also permitted individuals who did not have sufficient pension credits for a full pension to continue in employment. Some of these nations increased these incentives (the UK, Austria, France, Spain and Finland). Denmark and Greece created incentives for deferring pension receipt in the 2000s whilst Portugal and Luxembourg altered their rules to be more restrictive.

In addition, the EU has advocated the use of partial pensions to allow individuals to gradually retire and extend their working lives. This too was an area of policy divergence among EU15 nations. The UK, Ireland, Portugal, the Netherlands and Greece all had no options for combining part-time work with the receipt of a proportion of the pension whilst Sweden, Belgium, Finland, Spain, Germany, Austria, Italy, Denmark, France and Luxembourg conversely had partial pension options available. There was also a mix in terms of the reforms of these policies, where applicable. Some nations made these policies more accessible such as Sweden, Belgium, Finland and Spain whereas Germany retrenched their policy by raising the age threshold and Austria closed off its options for part-time pensions by 2010. In addition, as with decommodification options such as early exit and retirement, a number of these EU15 nations had contribution requirements for the receipt of partial pensions, including Denmark, Germany, Austria, Luxembourg, Spain, Italy and France (though Luxembourg and France also had more open options too). Sweden had opened up its partial pension policy by removing the contribution requirement.

This book therefore considered more than the retrenchment of early exit and retirement options *or* the creation of ALMPs and incentives for deferring pension receipt to take a holistic approach. The EU's active ageing agenda includes both the expansion of opportunities for labour market participation and the contraction of prospects for early withdrawal. A classification made by the EU (Council of the European Union, 2004) was returned to and updated to provide a framework to organise the empirical data on EU15 nations and also revealed some similarities in active ageing approaches. This revised classification was as follows:

- Group I: Nations that were in 2001 above or close to the Stockholm target and have since maintained or strengthened this position, including Sweden, Denmark, the UK, Ireland and Portugal.
- Group II: Nations that since 2001 have moved beyond the Stockholm target including the Netherlands, Finland and Germany.
- Group III: Nations that have not yet met the Stockholm target, but have made progress beyond the average percentage change for EU15 nations between 2001–10 (9.6%), including Belgium, Austria and Luxembourg.
- Group IV: Nations that have not yet met the Stockholm target and whose progress was less than the average for EU15 nations (9.6%), including France, Greece, Italy and Spain.

In Group I where the nations had exceeded or were close to the Stockholm target when it was established in 2001, there was a mix of those with no or limited decommodification opportunities but also narrow ALMPs (UK and Ireland) and others with more opportunities for labour market exit at the same time as policies for *de jure* recommodification (Sweden, Denmark and Portugal). However, compared to other EU15 nations, even those Group I countries with decommodification policies had far fewer than many nations in Groups II–IV. In terms of the reforms undertaken, these nations with early exit and retirement policies undertook significant retrenchment, in particular of policies with a rights-focus (Portugal, Ireland, Sweden and Denmark). For Portugal and Denmark, this meant the desert-element of their decommodification policies was strengthened resulting in a shift from fairly open systems to those that were strongly linked to labour market participation either in terms of contributions (or membership of certain occupations in the case of the former nation). However in Sweden and Ireland, rights-focused decommodification was the only option available in 1995 and therefore in these nations where decommodification was available to all, when these policies were closed off, all were equally subject to retrenchment.

Group II nations were all below the Stockholm target in 2002 but had all moved beyond it by 2010. The Netherlands, Finland and Germany all had decommodification options which had in common a mix of rights and desert as the eligibility criteria. These were significantly retrenched through the raising of age thresholds and in the case of Germany, restrictions for younger cohorts. Finland also promoted labour market participation in older ages by altering pension accrual arrangements. These nations, at the same time as narrowing opportunities for exit, promoted labour market participation through a broad range of ALMPs.

Whilst Group II nations had moved to surpass the Stockholm target, Groups III and IV failed to reach the 50% employment rate for those aged 55–64. Taking into account the different initial employment rates in 2001, Group III nations had progressed further than the average for all EU15 nations (+9.6% from 2001–10). These nations had in common a focus on incentivising the recruitment of older people through subsidies and social security exemptions for employers, whilst at the same time penalising those who made older workers redundant. They also undertook retrenchment of early exit and retirement policies through the increase of age thresholds and the imposition of reductions to final pension amounts, thus making them both more difficult to access and less appealing.

In Austria, significant but gradual retrenchment of the early retirement schemes began in the late 1990s which included the raising of age thresholds and contribution requirements as well as applying penalties which would reduce the final pension amount. At the same time, ALMPs were introduced with a strong focus on protecting older people in employment (including age discrimination legislation, an 'early warning' system to prevent redundancies through subsidies, flexible working-time practices, training for older workers and penalties for employers making older people redundant). Concurrently, there were 'carrots' for employers in the form of subsidies. Similarly, Belgium also from 2000 raised the contribution and age thresholds for early retirement as well as imposing penalties on those opting to leave the labour market early. In terms of ALMPs, in addition to job placements, Belgium also focused its attention on incentivising the recruitment of older people through subsidies. There was also a focus on the social partners with an agreement to improve the working and pay conditions of older workers. Luxembourg however did not engage in significant retrenchment of early retirement and exit options, or introduce a large number of ALMPs however those that were introduced did focus on incentivising the recruitment of older people.

Group IV nations included Spain, Greece, France and Italy. The latter had the furthest to travel to reach the 50% Stockholm target, having only 28% of those aged 55–64 in employment in 2001 whilst Spain had comparatively less ground to make up, with a rate of 39.2% in that year. Nonetheless all four of these nations had not reached the Stockholm target by 2010 and had also not increased their employment rate at the same pace as their EU15 counterparts (9.6%+ was the EU15 average) (Eurostat, 2011).

In terms of what these nations had in common in terms of policies and policy reform, Group IV nations all had a labour market element to their decommodification opportunities in that there were early exit and retirement policies for certain industries or with the requirement that the older person be replaced by a younger unemployed individual. Greece, France and Spain all had decommodification policies which included either or both contribution and industry-focused requirements whilst Italy included a more complex mix of both rights- and labour market-focused (with a contribution requirement) options. These nations all tended to focus retrenchment on the non-labour market-focused decommodification opportunities perhaps because agreements with social partners made these more difficult to reform than rights- or desert-focused policies. In this they are not dissimilar from the Group III nations in their focus on increasing the contribution requirements for desert-focused schemes while the labour market-orientated early retirement routes were relatively unreformed. Greece and Italy introduced more restrictive pension and early retirement schemes for younger cohorts, and perhaps therefore large gains in the participation rates of older workers will be seen among future generations. ALMPs also tended to focus on subsidies and protecting the employment of older workers.

Just as there was complexity at the macro-level, the policy picture at the micro-level in terms of impact of reform on policy choices available to sub-groups within the 55–64 cohort was also far from simple. Even within individual EU15 nations there is divergence with regard to the policy treatment of different groups within the older age cohort, with characteristics such as age, gender, occupation and employment history impacting upon individual choice. This book used the following model biographies to explore the different policy treatment of different groups within the older age cohort:

– the 50-plus:
  – *Laurent*: Aged 55 with 35 years of employment contributions.
  – *Jean*: Aged 55 with a disjointed employment history.

- the 60-plus:
  - *Sasha*: Aged 63 years with 35 years of employment contributions.
  - *Jude*: Aged 63 with a disjointed employment history.

The assumption of the recommodification of labour argument is that policies for the decommodification of labour have been replaced by policies for its recommodification. Similarly, the reserve army of labour literature would suggest that older people as a group are pushed into and out of the labour market when the economy requires. However, as the data from the model biographies demonstrates, it was more often the case that decommodification policies were not available to *all* older people in the first place – these policies' eligibility criteria included elements such as contribution records and occupation and so excluded as much as they included older individuals in the decommodification or reserve army of labour. Equally, the reserve and recommodification of labour arguments would imply that *all* older people are being pulled into the labour market but again the data demonstrate that some subgroups within this age group retained the ability to be decommodified whilst their peers were increasingly being encouraged to re-engage with or remain in the labour market. However, it was clear in most of the EU15 nations that the policies available which allowed older people to exit and retire early were being retrenched over the period from the mid-1990s to 2010. The impact on the model biographies varied according to the manner of retrenchment undertaken.

The three main characteristics which delineated the model biographies' hypothetical access to policies for decommodification and thus made them less subject to recommodification are age, gender and labour market history. In terms of the latter, related to the identification of desert-focused decommodification options, individual contribution records were in many EU15 nations key to accessing early exit and retirement options in 1995 (Germany, Austria, Denmark, the Netherlands, Greece, France, Italy, Spain, Belgium, Luxembourg and Finland), and became increasingly important over time as more universal policies were retrenched. Nations with a desert-focus to their eligibility criteria in addition to universal options closed off the latter whilst also restricting access to the former through increased contribution requirements, age thresholds and penalties for early retirement (for example Denmark, Germany and the Netherlands). The focus on desert as a means of deciding whether individuals can exit or retire early raises issues related to intragenerational equity. Perhaps those who have worked for the longest deserve to leave the labour market earlier and not age as actively (in accordance with the EU's narrow focus on employment). However, this can only be considered equitable if employment opportunities are genuinely equally distributed and an individual's gender, caring responsibilities, race, age and other characteristics do not interfere with their labour market participation.

Labour market history was also pertinent in terms of previous occupations. In nations where early retirement options were limited to certain types of employment, these policies were largely retained (France, Italy, Spain, Luxembourg and Belgium) whilst desert-focused routes were retrenched through increased contribution requirements, age thresholds or the application of pension reductions for those exiting early. Again, perhaps allowing early exit and retirement for those in certain industries reflects the fact that some jobs are demanding and therefore early withdrawal from the labour market is necessary. Yet on the other hand, these policies are potentially a tool for employers to shed older workers who are perhaps the most expensive in terms of wage costs. Either way, these policies run the risk of creating divisions within the cohort of 'older age' where some are required to age more actively than others, depending on their occupation.

With regard to age, this book has focused on policies for 'older' people, i.e. those aged 55–64. What became apparent is that the policy options for those within this cohort are not the same. Typically those closer to state pension ages had more opportunities to retire prematurely yet because there was variation in these standard retirement ages among EU15 nations, there too were differences in how 'early' early retirement could be. For example, France had a comparatively low state pension age of 60, which in countries such as Sweden and Germany was the threshold for early as opposed to conventional retirement. Age also became increasingly important in determining decommodification options as the raising of age thresholds was often the way early retirement and exit policies were retrenched (for example in Sweden, Portugal, Finland, Germany, Austria, Spain and Italy).

Age was also important in terms of cohort membership in Austria, Germany, Greece and Italy. All of these nations had different systems in place for different birth cohorts, with those available to younger people more restrictive in terms of exit and more focused on incentivising longer working lives. Age and gender also intersect in a number of countries where male and female pension ages (including for early retirement and exit) were not harmonised, including the UK, Austria, Greece, Italy (for those under the *Amato* reform), Portugal and Belgium though the latter two began to harmonise their pension ages by 2010.

Aside from the different age thresholds for men and women, gender also impacted on the model biographies' access to decommodification options in terms of labour market contributions. As noted in Section 3.3, the degree to which care is provided outside of the family (either by the state, the market or third sector) influences the amount which is left to the family, primarily by women. It is argued therefore that the state's provision of care can assist women in entering and remaining in paid employment and therefore accruing sufficient pension credits to retire or exit earlier than state pension age.

In brief therefore, the conclusions of this book are threefold. First, at the macro-level, between the mid-1990s and 2010 there has indeed been a shift towards the EU's active ageing agenda and thus older individuals' labour is being recommodified and yet national reforms and final policy mixes still contained a great deal of divergence. Thus to present an overly convergent or divergent picture would be misleading: progress has been made in all EU15 nations towards the EU's active ageing agenda, but this does not mean that by 2010 they presented the same – or even very similar – policy packages for older individuals. The recommodification of labour argument too readily assumes nations are adopting similar workfare/activation-based employment policies, and indeed, that at one point, decommodification was widely available. The data demonstrate that the decommodification policies available in EU15 nations were diverse both in terms of their number and eligibility criteria. At the micro-level, the recommodification of labour argument firstly implies that all individuals had the same capacity for decommodification, and that all are equally subject to policies that focus on labour market participation. Both at the national and individual level, the data demonstrate decommodification and recommodification policies were implemented and experienced differentially.

However, though the recommodification of labour argument needs to be refined with respect to the policy treatment of older individuals, it is clear from the data that EU15 nations have all made some progress towards the goals of active ageing, as outlined by the EU and could be seen indicative of the decline of decommodification and the ascendancy of recommodification approaches. In addition, as highlighted in Chapters 4–7, these nations had very different policy legacies in terms of decommodification options, which not only provided the parameters for reform, but they also dictated the distance nations had to travel towards active ageing approaches. Therefore EU15 nations' diverse policy approaches to those over 50 reflect their different starting points in terms of the legacy of existing decommodifying policies. The nature and speed of retrenchment and reforms, although largely towards the aims of the active ageing agenda (to retain and reintegrate older workers in the labour market), varied and to some degree can be grouped in accordance with the original decommodification approach adopted by nations at the initial point of the enquiry. As a result, and due to the dynamic nature of policy, nations may continue to converge upon the EU ideal of active ageing and increase the recommodification of older individuals' labour. Though this book attempted a methodological approach that captured change, policy has continued to progress beyond 2010 and nations could now embody a more unified recommodification approach, or may in the future.

Second and related, the distance nations had to travel towards these goals as a result of pre-existing decommodification policy legacies affected the speed, direction of reform and final policy mixes. For example, nations with desert-focused decommodification could increase contribution requirements whilst those without a desert-element could introduce insurance thresholds, thereby shifting the focus of their early exit and retirement approach.

Finally, at the micro-level, the model biographies assisted in providing more nuanced accounts of the recommodification and reserve army of labour literature. It is clear that from the outset, policies in certain nations divided individuals into those worthy of decommodification and those who were not. As a result, the notion that *all* individuals are being recommodified and *all* older individuals are part of the reserve army of labour was refined as clearly some had more limited opportunities for labour market exit at the outset. The data demonstrate divergence with respect to national policy treatment of the older age cohort along the lines of age, gender, occupation and previous contributions. At the individual level, active ageing policies have distinguished between different groups within the older age cohort, recommodifying some, whilst other retain the ability to be decommodified. Not all older individuals were able to be decommodified when the labour market did not require their presence, and similarly, in light of the dire warnings regarding demographic ageing and replacement rates, not all are being pulled back into the labour market. Divergence exists between different ages, genders, occupations, work histories and birth cohorts. Thus the notion that *all* older individuals make up the reserve army of labour, pulled into the labour market when there is a labour shortage needs to be evaluated. The political economy of ageing arguments around the importance of policy and labour market experience in shaping the experience of old age have been borne out by the data. Policy dictates whether ageing should be active or passive, and clearly not all individuals are equally subject to the demands of the market. Thus convergence towards the EU vision of active ageing and the recommodification of labour is more complex, with nations adopting a variety of different reforms and policy mixes, which in turn focus on different groups within the older age cohort.

Notes

Chapter 1   Introduction

1   'Older' in this volume refers to individuals aged 50–74, though this definition is debated. Bytheway (1995) argues it is defined statically as those aged 50–74, whereas the European Union defines older workers as either as those aged 45–64 in its documents (Committee of the Regions, 2003) or as those aged 55–64 when setting targets for its member states (e.g. those of Barcelona and Stockholm which aim for the extension of the effective retirement age by five years and 50% employment by 55–64 year olds in member states). Researchers, Taylor (2006: 3) argues "tend to view people aged 50 years and over as 'older' based on a sharp decline in labour force participation rates after this age. However, it is clear that what is defined as 'old' varies markedly between industrial sectors and occupational groups. It is also sometimes argued that women are considered by employers as 'old' at younger ages than men". This book takes a synthesis of these perspectives to focus on policies in EU15 nations aimed at those aged 50–74, as this allowed for the examination of both specialist reforms aimed at reintegrating unemployed individuals over the age of 50, as well as those designed to encourage individuals to work beyond the state pension age, thus reflecting the Barcelona and Stockholm targets. Data from the Eurobarometer survey also found that individuals themselves favour either this term or 'senior citizens' over 'third age' (Walker and Maltby, 1997).

2   The Stockholm (2001) and Barcelona (2002) targets which respectively aim for 50% employment rates for older individuals and the extension of the effective retirement age by five years.

3   The rationale for the focus on these policy areas will be explained in Section 2.1.

Chapter 2   Active Ageing: Origins and Resurgence

1   The WHO is not alone in promoting a holistic approach to active ageing. Walker (2002) advocates a lifecourse approach, focusing on all areas of life, as opposed to labour market participation.

2   Equally, these organisations are not the only proponents of this approach. The Transitional Labour Market Theory (TLM) argues individuals should not be forced or enticed out of the labour market and active ageing should provide the opportunity for individuals to retrain and access new and fulfilling jobs. At the same time, flexible working options should be made available to allow individuals to work less hours or in less demanding positions (Hartlapp and Schmind, 2008: 411).

3   This concept has its origins in the work of Karl Polanyi (1944) who argued the post-industrial state and economy are intertwined as an integral feature

of capitalism includes the state's intervention in the market to secure conditions for profit; something the market itself is unable to procure (c.f. Block and Evans, 2005; Jessop, 1999; also related is Marx's concept of the 'conditions of production' (1973)). He argued the industrialisation of the 19th century prompted a departure from the traditional structure whereby the economic sphere was a function of the social sphere. Indeed, what took place was a role reversal with the social subordinated to the needs of the economy as "[a] market economy can only exist in a market society" (Polanyi, 1944: 71) and land, labour and money were created and maintained by the state as 'fictitious commodities' (in that their value cannot be separated from their form). However, Polanyi posited that a real self-regulating market is not conducive to society's survival in that "[s]uch an institution could not exist for any length of time without annihilating the human and natural substance of society; it would have physically destroyed man and transformed his surroundings to wilderness" (Polanyi, 1944: 3). Society's natural response is to protect itself from the ravages of the self-regulating market yet with such intervention the self-regulating market ceases to be such. Indeed, as workers compete as commodities, the price of labour depreciates and thus a system of decommodification is required to safeguard against pauperism.

4   However across EU nations there is variation with Portugal and Sweden seeing a rise in the active population between 1995 and 2001 (+1.0% and +1.3% respectively), whilst Italy (–7.5%), Germany and Spain (both –6.6%) have seen a decline (Committee of the Regions, 2003).

5   Germany includes ex-GDR from 1991. France is not included as the data were unavailable.

6   The dependency ratio is the ratio between the total number of older persons of an age when they are generally economically inactive (aged 65 and over) and the number of persons of working age (from 15 to 64).

7   When referring to the 'social contract', Walker is referring to the "social policy contract based on intercohort transfers of resources through the mediums of taxation and social expenditure" (Walker, 1996b: 13).

8   The service sector itself represents the tertiary sector after the primary sector concerned with the extraction of raw materials and the secondary sector focused on manufacturing (Clark, 1957).

Chapter 3   The EU's Active Ageing Agenda

1   http://europa.eu/abc/panorama/index_en.htm

2   An example of this argument is Ney's (2005) work which focuses specifically on national approaches to active ageing, which he argues, are mediated by countries' institutional contexts as new policies are actualised in the wake of existing schemes which make certain options less politically and practically feasible. Thus Ney argues Nordic states are characterised by universal, generous benefits, which is reflected in their active ageing approach. However, these nations also focus on active ageing through training. Active ageing in the 'North-Western Fringe' including the UK is encompassed within the broader 'less eligibility' approach and therefore aims to remove barriers to labour market participation including age discrimination. Continental Europe, Ney argues is concerned

with ensuring their welfare states are modified to allow them to compete in a global market and therefore are concerned with creating flexibility at the end of individuals' working lives and retraining where beneficial to the market.

3 Miller (1989) uses the example of a man who is a Grade C official and therefore has the right to a salary of £3,000 by virtue of this status. Miller argues if the salary is contingent on the man having been particularly hardworking, this is a question of desert. Thus rights are derived through status; desert is earned through behaviour and is "a backward-looking concept" (Miller, 1989: 93).

4 With regard to the Barcelona target of extending working lives by five years, nations have not made as much progress and the available data tends to be more limited than for employment rates. Thus for the purpose of this volume, nations will be explored using a typology based on the CEU's 2004 report and subsequent progress towards the Stockholm target. This typology provided a framework for presenting the data; the author acknowledges that factors other than social policies may have caused EU15 nations to possess similar employment rates for older people.

5 The timeframe selected is from the mid-1990s to 2010. This is in order to examine the policy change from a time pre-the resurgence of the active ageing agenda up until the deadline for the Stockholm target (the first reference to active ageing was in 1999 (Commission of the European Communities, 1999: 6)). The distance nations had to travel towards active ageing policies was therefore also taken into account.

Chapter 4 Group I: The Vanguards – Consolidating Their Position Beyond Stockholm

1 A disability pension could still be accessed by all from the age of 17 and thus will not be included further in this volume as the focus is on policies available to those over the age of 50.

2 In addition, there was a disability pension scheme open to those aged 18 to 66 and though its claimants were predominantly older, Hansen (2002) argues it cannot be seen as an early exit route in the strictest sense.

Chapter 5 Group II: Surpassing Stockholm

1 http://www.bertelsmann-stiftung.de/cps/rde/xchg/SID-0A000F14-A86517EC/bst_engl/hs.xsl/54224_54228.htm?reform.id=67949.xml&rm.show=reform

2 By 1993, the classical medical criteria for disability benefit (WAO) were expanded to include the consideration of illness in the context of the local labour market. As a result, "[p]artial invalidity may thus be recognised as total invalidity and entitlement to a full pension" (Guillemard, 1993: 39). Thus invalidity became the main impetus for early exit.

3 The VUT system was reformed in 1997 with the Pension Covenant whereby the government and social partners altered the funding to represent a more individualised scheme which meant the benefits received would be directly linked to the contributions made over the individual's lifecourse. Though

the state had no legal basis to intervene in the VUT schemes, they could do so indirectly through fiscal policy (OECD, 2005j; de Vroom, 2004b: 143). Nonetheless, they cannot be seen as a state-provided early retirement.

4 In terms of other non-state early exit routes, the pay-as-you-go early retirement scheme, the VUT, was gradually being replaced by an occupational pre-pension scheme, based on an individual life savings model (OECD, 2005j). This new system only applied to those under 55 in 2005. The individual could exit using the lifecourse savings scheme which provided benefits for three years at 70% of their previous earnings (OECD, 2005j).

5 http://www.bertelsmann-stiftung.de/cps/rde/xchg/SID-0A000F14-A86517EC/bst_engl/hs.xsl/54224_54228.htm?reform.id=67406.xml&rm.show=reform

6 The disability pension (*Työkyvyttömyyseläke*) has not been included here as it is open to those aged 16–64.

7 There was also a disability pension (*Erwerbsunfähigkeitsrenten*), accessible regardless of age for those who had five insurance years, three of the previous five years in activity and a medical condition that reduced their working capacity. As with Denmark and in line with Hansen's (2002) arguments, this cannot be considered to be an early retirement route as it was open to all ages. Indeed, as Börsch-Supan and Jürges (2011) note, the old age disability pension explored above focused largely on age as its eligibility criteria whilst the disability pensions focused mainly on work capacity, though they note that in practice this is less precise. The benefit was to become a key means of early exit however, with the number of 55–59 year olds claiming the benefit increasing from 26,000 in 1975 to 56,000 in 1984 (Mandin, 2004).

8 However, the new German constitution removed the right to work which had been part of the German Democratic Republic (GDR)'s constitution. The onus is instead placed on workers' committees to make sure there are no age limits (Taylor, 1998).

9 This Act was removed due to its costs for the state and perhaps because the dual aims of creating vacancies for younger unemployed individuals whilst dissuading employers from making older individuals redundant presented a paradox.

Chapter 6   Group III: Below Stockholm but Approaching Fast

1 Belgium is characterised by regional differences between Flanders, Wallonia and Brussels in terms of policies. This section however addresses the policies implemented on a national scale.

2 The Disablement Insurance Act is not included here as it is open to all ages.

3 In addition to the main routes, there were also 'Canada Dry' schemes, so named because they look "like a pension, but is not a pension. This is reminiscent of an old commercial for Canada Dry ginger ale on the European continent, according to which it had the color of beer and the taste of beer, but it was not beer" (Cremer et al., 2009). The 'Canada Dry' schemes allowed employers to dismiss older workers who would then receive unemployment benefits plus an additional supplement paid by their former firm. This scheme, Taylor (2005: 11) argues "acts as a form of early retirement but avoids the usual legal con-

straints and places most of the cost on the state's unemployment funds. It is usually granted to older employees who do not fulfil the legal requirements for a conventional pre-pension".

Chapter 7   Group IV: The Laggards – Slow Progress Towards Stockholm

1 Unlike the Dutch VUT scheme, though the ARPE was managed by the social partners, all negotiations and suggested reforms required the consent of the state. Indeed, all meetings of the UNEDIC (comprising of five seats each for the main employer and employee unions) are overseen by a representative of the state (*Contrôleur d'Etat*) as a non-voting member. UNEDIC schemes were financed by employee and employer contributions, but also receive state subsidies from the Ministry of Labour when required in instances of deficit. Daley (1996) argues the latter gives the state leverage in negotiations around reforms.
2 There was a similar full-time scheme created in the same year which ceased to accept new applicants from 1984 (solidarity pre-retirement contracts, *Allocation conventionnelle de solidarité* (ACS)) (Guillemard, 1991).

Chapter 8   Policy Convergence, Divergence and Intragenerational Equity in EU15 Nations

1 The UK represents an outlier in that there are no early exit or retirement schemes available, and there have not been since the Job Release Scheme was abolished in 1989.
2 I use the term 'tier' in line with Immergut and Anderson (2007)'s distinction whereby a 'pillar' indicates the sector in which the pension scheme is located (public, occupational and private) and 'tiers' include the type of benefit. Thus 'first tier' includes basic state pensions, whilst social assistance substitutes for a minimum pension are separate.
3 However, Sweden had introduced a new pension system by 2001, with individuals subject to different ratios of old and new system depending on the year of their birth. The new system was made up of three elements, including earnings-related and insurance-based strands but retained a universal basic scheme.

Chapter 9   Conclusion

1 Increasing the participation rates of those aged over 55 to 50% and extending the effective retirement age by five years respectively.

Bibliography

Amitsis, G., Berghans, J., Hemerijck, A., Sakellaropoulos, T., Stergiou, A. and Stecvens, Y. (2003) *Connecting Welfare Diversity within the European Social Model*, Background report for the International Conference of the Hellenic Presidency of the European Union on the Modernisation of the European Social Model (May), Ioannina, Greece.

Anderson, E. (2004) 'Welfare, Work Requirements, and Dependant-Care', *Journal of Applied Philosophy*, Vol. 21 (3): 243–247.

Anderson, K. and Immergut, E. (2007) 'Sweden: After Social Democratic Hegemony', in Immergut, E., Anderson, K. and Schulze, I. (eds) *The Handbook of West European Pension Politics*, Oxford: Oxford University Press.

Anttonen, A. and Sipilä, J. (1996) 'European Social Care Services: Is It Possible to Identify Models?', *Journal of European Social Policy*, Vol. 6: 87–100.

Araico, A. S. (2004) 'Ageing and Work in Spain: The End of Working Life?' in De Vroom, B., Maltby, B., Mirabile, M. L. and Øverbye, E. (eds) *Ageing and the Transition to Retirement: A Comparative Analysis of European Welfare States*, Aldershot: Ashgate.

Arber, S. (2006) 'Gender and Later Life: Change, Choice and Constraints', in Vincent, J., Phillipson, C. and Downs, M. (eds) *The Futures of Old Age*, London: Sage Publications.

Argoud, D. and Guillemard, A. (1999) 'The Politics of Old Age in France', in Naegele, G. and Walker, A. *The Politics of Old Age in Europe*, Buckingham: Open University Press.

Arts, W. and Gelissen, J. (2002) 'Three Worlds of Welfare Capitalism or More? A State-of-the-Art Report', *Journal of European Social Policy*, Vol. 12 (2): 137–158.

Attas, D. and De-Shalit, A. (2004) 'Workfare: The Subjection of Labour', *Journal of Applied Philosophy*, Vol. 21 (3): 310–322.

Avramov, D. (2003) 'Chapter 1 – Active Ageing: Setting the Stage', in Avramov, D. and Maskova, M. *Active Ageing in Europe: Volume 1*, Strasbourg: Council of Europe Publishing.

Avramov, D. and Maskova, M. (2003) *Active Ageing in Europe: Volume 1*, Strasbourg: Council of Europe Publishing.

Barbier, J-C. and Théret, B. (2000) *The French Social Protection System: Path Dependencies and Societal Coherence*, The Year 2000 International Research Conference on Social Security, Helsinki, 25–27 September 2000.

Barros, A. and Coelho, P. (1998) 'Portugal', in Taylor, P. for the Employment and European Social Fund *Projects Assisting Older Workers in European Countries: A Review of the Findings of Eurowork Age*, Luxembourg: Office for Official Publications of the European Communities.

Belgium (1999) *Investing in People and Jobs: Belgium's 1999 National Action Plan Prepared in Accordance with European Employment Guidelines*: http://ec.europa.eu/employment_social/employment_strategy/nap_1999/napbe_en.pdf, date accessed 14/04/09.

Belgium (2002) *Plan d'action belge 2002 établi dans le cadre des lignes directrices européennes pour l'emploi*: http://ec.europa.eu/employment_social/employment_strategy/nap_2001/napbe_fr.pdf, date accessed 14/04/09.

Belgium (2003) *European Strategy for Employment 2003: National Action Plan for Employment – Belgium*: http://ec.europa.eu/employment_social/employment_strategy/nap_2003/nap_be_fr.pdf, date accessed 14/04/09.

Belloni, M., Monticone, C., Trucchi, S. and Fornero, E. (2006) *Flexibility in Retirement. A Framework for the Analysis and a Survey of European Countries*, Report realized by CeRP for the DG Employment, Social Affairs and Equal Opportunities, European Commission; October 2006: http://servizi.econ.unito.it/cerp/site/publications/flexibility_in_retir, date accessed 27/10/2008.

Bengtson, V., Putney, N. and Johnson, M. (2005) 'The Problem of Theory in Gerontology Today', in Johnson, M. *The Cambridge Handbook of Age and Ageing*, Cambridge: Cambridge University Press.

Bentoumi, M. and Evans, A. (2008) *2008 Social Security Funding Law (LFSS 2008): Impact on Corporate Accounting in France*, http://www.watsonwyatt.com/news/featured/pdf/2008_Social_Security_Funding_Law.pdf

Bertelsmann-Stiftung Foundation (1999) *Reform: Reform of the Public Employment Services (PES) (Italy)*, http://www.bertelsmann-stiftung.de/cps/rde/xchg/SID-0A000-F14-A86517EC/bst_engl/hs.xsl/54224_54228.htm?reform.id=63335.xml&rm.show=reform, date accessed 27/10/2008.

Bertelsmann-Stiftung Foundation (2000a) *Reform: Pension Reform – Raising Actual Retirement Age (Austria)*, http://www.bertelsmann-stiftung.de/cps/rde/xchg/SID-0A000F14-F1AE2DB8/bst_engl/hs.xsl/54224_54228.htm?reform.id=64308.xml&rm.show=reform, date accessed 27/10/2008.

Bertelsmann-Stiftung Foundation (2000b) *Reform: Reform of Disability Pension (Denmark)*, http://www.bertelsmann-stiftung.de/cps/rde/xchg/SID-0A000F14-F1AE2DB8/bst_engl/hs.xsl/54224_54228.htm?reform.id=64403.xml&rm.show=reform, date accessed 27/10/2008.

Bertelsmann-Stiftung Foundation (2001a) *Reform: Employment Premium (Prime pour l'emploi) (France)*, http://www.bertelsmann-stiftung.de/cps/rde/xchg/SID-0A000F14-F1AE2DB8/bst_engl/hs.xsl/54224_54228.htm?reform.id=65535.xml&rm.show=reform, date accessed 27/10/2008.

Bertelsmann-Stiftung Foundation (2001b) *Reform: Reform of the Unemployment Insurance Scheme (France)*, http://www.bertelsmann-stiftung.de/cps/rde/xchg/SID-0A000F14-F1AE2DB8/bst_engl/hs.xsl/54224_54228.htm?reform.id=64443.xml&rm.show=reform, date accessed 27/10/2008.

Bertelsmann-Stiftung Foundation (2002a) *Reform: Reintroduction of Pension for Bereaved (Denmark)*, http://www.bertelsmann-stiftung.de/cps/rde/xchg/SID-0A000F14-F1AE2DB8/bst_engl/hs.xsl/54224_54228.htm?reform.id=66080.xml&rm.show=reform, date accessed 27/10/2008.

Bertelsmann-Stiftung Foundation (2002b) *Reform: The Special Pension Savings (Denmark)*, http://www.bertelsmann-stiftung.de/cps/rde/xchg/SID-0A000F14-F1AE2DB8/bst_engl/hs.xsl/54224_54228.htm?reform.id=66069.xml&rm.show=reform, date accessed 27/10/2008.

Bertelsmann-Stiftung Foundation (2002c) *Reform: Activation Offers to Certain Unemployed Members of Unemployment Insurance Funds (Denmark)*, http://www.bertelsmann-stiftung.de/cps/rde/xchg/SID-0A000F14-F1AE2DB8/bst_engl/hs.xsl/ 54224_54228.htm?reform.id=66115.xml&rm.show=reform, date accessed 27/10/2008.

Bertelsmann-Stiftung Foundation (2003a) *Reform: Agenda 2010 (Germany)*, http://www.bertelsmann-stiftung.de/cps/rde/xchg/SID-0A000F14-F1AE2DB8/bst_engl/

hs.xsl/ 54224_54228.htm?reform.id=67351.xml&rm.show=reform, date accessed 27/10/2008.

Bertelsmann-Stiftung Foundation (2003b) *Reform: Proposed Act to Strengthen the Right to Full-Time Employment (Sweden)*, http://www.bertelsmann-stiftung.de/cps/rde/ xchg/SID-0A000F14-A86517EC/bst_engl/hs.xsl/54224_54228.htm?reform.id=68060. xml&rm.show=reform, date accessed 27/10/2008.

Bertelsmann-Stiftung Foundation (2003c) *Reform: The Working Life Focussed Rehabilitation Process (Sweden)*, http://www.bertelsmann-stiftung.de/cps/rde/xchg/SID-0A000F14-A86517EC/bst_engl/hs.xsl/54224_54228.htm?reform.id=64698.xml&rm.show=reform, date accessed 27/10/2008.

Bertelsmann-Stiftung Foundation (2003d) *Reform: Reforms in the Unemployment Insurance System (Sweden)*, http://www.bertelsmann-stiftung.de/cps/rde/xchg/SID-0A000F14A86517EC/bst_engl/hs.xsl/54224_54228.htm?reform.id=64186.xml&rm.show=reform, date accessed 27/10/2008.

Bertelsmann-Stiftung Foundation (2004a) *Reform: Pensions Act 2004 (United Kingdom)*, http://www.bertelsmann-stiftung.de/cps/rde/xchg/SID-0A000F14-F1AE2DB8/bst_engl/hs.xsl/54224_54228.htm?reform.id=68672.xml&rm.show=reform, date accessed 27/10/2008.

Bertelsmann-Stiftung Foundation (2004b) *Reform: Expected Reform of the Taxation of Public Pensions in Germany (Germany)*, http://www.bertelsmann-stiftung.de/cps/rde/ xchg/SID-0A000F14-E925D852/bst_engl/hs.xsl/54224_54228.htm?reform.id= 66155.xml&rm.show=reform, date accessed 27/10/2008.

Bertelsmann-Stiftung Foundation (2004c) *Reform: Pension Reform August 2004 (Italy)*, http://www.bertelsmann-stiftung.de/cps/rde/xchg/SID-0A000F14-A86517EC/bst_engl/hs.xsl/54224_54228.htm?reform.id=68380.xml&rm.show=reform, date accessed 27/10/2008.

Bertelsmann-Stiftung Foundation (2005a) *Reform: Abolishment of Tax Advantages for Early Retirement through VUT and Pre-pension Systems (Netherlands)*, http://www. bertelsmann-stiftung.de/cps/rde/xchg/SID-0A000F14-A86517EC/bst_engl/ hs.xsl/ 54224_54228.htm?reform.id=67949.xml&rm.show=reform, date accessed 27/10/2008.

Bertelsmann-Stiftung Foundation (2005b) *Reform: Harmonisation of the Austrian Pension System (Austria)*, http://www.bertelsmann-stiftung.de/cps/rde/xchg/SID-0A000F14-F1AE2DB8/bst_engl/hs.xsl/54224_54228.htm?reform.id=68439. xml&rm.show=reform, date accessed 27/10/2008.

Bertelsmann-Stiftung Foundation (2010) *International Reform Monitor 2010: The Netherlands*, http://www.bertelsmann-stiftung.de/cps/rde/xbcr/SID-C17805C9-362C9ABF/bst_engl/Netherlands_Topics.pdf, date accessed 23/07/2011.

Bettio, F. and Plantenga, J. (2004) 'Comparing Care Regimes in Europe', *Feminist Economics*, Vol. 10 (1): 85–113.

Billari, F. and Galasso, V. (2008) *What Explains Fertility? Evidence from Italian Pension Reforms*, Working Papers 343, IGIER (Innocenzo Gasparini Institute for Economic Research): Bocconi University.

Block, F. and Evans, P. (2005) 'The State and the Economy', in Smelser, N. and Swedberg, R. (eds) *The Handbook of Economic Sociology*, Princeton and Oxford: Princeton University Press.

Boldrin, M., Jimenez-Martin, S. and Peracchi, F. (1999) 'Social Security and Retirement in Spain', in Gruber, J. and Wise, D. *Social Security and Retirement Around the World*, Chicago: Chicago University Press.

Bond, J. (1986) 'Political Economy as a Perspective in the Analysis of the Experience of Old Age', in Phillipson, C., Bernard, M. and Strang, P. *Dependency and Interdependency in Old Age – Theoretical Perspectives and Policy Alternatives*, London: Croom Helm.

Bonoli, G. (2010) *The Political Economy of Active Labour Market Policy*, Working Papers on the Reconciliation of Work and Welfare in Europe, REC-WP 01/2010: http://www.socialpolicy.ed.ac.uk/_data/assets/pdf_file/0010/39268/REC-WP_0110_Bonoli.pdf, date accessed on 17/07/10.

Börsch-Supan, A. and Jürges, H. (2011) *Disability, Pension Reform and Early Retirement in Germany*, NBER Working Paper No. 17079.

Bradshaw, J., Ditch, J., Holmes, H. and Whiteford, P. (1993) 'Child Support in Fifteen Countries', *Journal of European Social Policy*, Vol. 3 (4): 255–271.

Bredgaard, T. and Tros, F. (2006) *Alternatives to Early Retirement? Flexibility and Security for Older Workers in the Netherlands, Denmark, Germany and Belgium*. Paper presented at the ILP Innovating Labour Market Policies, Amsterdam, 30 November and 1 December.

Bytheway, B. (1995) *Ageism*, Buckingham: Open University Press.

Carmel, E., Hamblin, K. and Papadopoulos, T. (2007) 'Governing the Activation of Older Workers in the European Union: The Construction of the "Activated Retiree"', *International Journal of Sociology and Social Policy*, Vol. 27 (9/10): 387–400.

Carmel, E. and Papadopoulos, T. (2003) 'The New Governance of Social Security in the UK', in Millar, J. (ed.) *Understanding Social Security*, Bristol: Policy Press.

Caro, F., Bass, S. and Chen, Y-P. (1993) 'Introduction: Achieving a Productive Aging Society', in Bass, S., Caro, F. and Chen, Y-P. (eds) *Achieving a Productive Aging Society*, London: Auburn House.

Carr, D. and Sheridan, J. (2001) 'Family Turning-Points and Career Transitions at Midlife', in Marshall, V., Heinz, W., Krüger, H. and Verma, A. (eds) *Restructuring Work and the Life Course*, Toronto: University of Toronto Press.

Casey, B. (1998) *Incentives and Disincentives to Early and Late Retirement*, OECD, Geneva, Working Paper AWP 3.3.

CEC (Commission of the European Communities) (1999) *Towards a Europe for All Ages – Promoting Prosperity and Intergenerational Solidarity*, COM(1999) 221 final: http://ec.europa.eu/employment_social/social_situation/docs/com221_en.pdf, date accessed on 11/07/05.

CEC (2002) *Europe's Response to World Ageing: Promoting Economic and Social Progress in an Ageing World. A Contribution of the European Commission to the 2nd World Assembly on Ageing*, COM(2002) 143 final: http://europa.eu.int/comm/employment_social/social_situation/docs/com2002_0143_en.pdf, date accessed 24/06/05.

CEC (2004) *Communication from the Commission to the Council, the European Parliament, the European Economic and Social Committee and the Committee of the Regions: Increasing the Employment of Older Workers and Delaying the Exit from the Labour Market*, Brussels, 3.3.2004, COM(2004) 146 final, http://eur-lex.europa.eu/LexUriServ/LexUriServ.do?uri=COM:2004:0146:FIN:EN:PDF, date accessed 26/03/12.

CEC (2005) *Green Paper 'Confronting Demographic Change: A New Solidarity Between the Generations'*, Brussels, 16.3.2005 COM(2005) 94 final: http://europa.eu.int/comm/employment_social/news/2005/mar/comm2005-94_en.pdf, date accessed 24/06/05.

CEU (Council of the European Union) (2000) *Council Directive 2000/78/EC: Establishing a General Framework for Equal Treatment in Employment and Occupation,* http://europa.eu.int/comm/employment_social/fundamental_rights/pdf/legisln/2000_78_en.pdf, date accessed on 11/11/05.

CEU (2004) *Draft Joint Employment Report 2003/2004,* 7069/04: http://ue.eu.int/ueDocs/cms_Data/docs/pressdata/en/misc/79690.pdf, date accessed 24/05/07.

Chuliá, E. (2007) 'Spain: Between Majority Rule and Incrementalism', in Immergut, E., Anderson, K. and Schulze, I. (eds) *The Handbook of West European Pension Politics,* Oxford: Oxford University Press.

Chuliá, E. and Asensio, M. (2007) 'Portugal: In Search of a Stable Framework', in Immergut, E., Anderson, K. and Schulze, I. (eds) *The Handbook of West European Pension Politics,* Oxford: Oxford University Press.

Clark, C. (1957) *The Conditions of Economic Progress,* 3rd ed., London: Macmillan.

COR (Committee of the Regions) (2003) *Older People in Working Life: Presentation of Relevant Actions by Local and Regional Authorities,* Luxembourg: Office for Official Publications of the European Communities.

Cremer, H., Lozachmeur, J. M. and Pestieau, P. (2009) 'Use and Misuse of Unemployment Benefits for Early Retirement', *European Journal of Political Economy,* Vol. 27 (2): 174–185.

Crompton, R., Dennett, J. and Wigfield, A. (2003) *Organisations, Careers and Caring,* Bristol: The Policy Press for the Joseph Rowntree Foundation, http://www.jrf.org.uk/sites/files/jrf/jr125-carers-employment-policies.pdf, date accessed 24/06/07.

Cumming, E. and Henry, W. (1961) *Growing Old: The Process of Disengagement,* New York: Basic Books.

Daley, A. (1996) *Steel, State, and Labor: Mobilization and Adjustment in France,* Pittsburgh: University of Pittsburgh Press.

de Vroom, B. and Blomsma, M. (1991) 'The Netherlands: An Extreme Case', in Kohli, M., Rein, M., Guillemard, A-M. and Gunsteren, H. (1991) *Time for Retirement: Comparative Studies of Early Exit from the Labor Force,* Cambridge: Cambridge University Press.

de Vroom, B. (2004a) 'Age Arrangements, Age Culture and Social Citizenship: A Conceptual Framework for Institutional and Social Analysis', in de Vroom, B., Maltby, B., Mirabile, M. and Øverbye, E. *Ageing and Transitions into Retirement: A Comparative Analysis of European Welfare States,* Aldershot: Ashgate.

de Vroom, B. (2004b) 'The Shift from Early to Late Exit: Changing Institutional Conditions and Individual Preferences: The Case of The Netherlands', in de Vroom, B., Maltby, B., Mirabile, M. and Øverbye, E. *Ageing and Transitions into Retirement: A Comparative Analysis of European Welfare States,* Aldershot: Ashgate.

Dean, H. (2007) 'The Ethics of Welfare-To-Work', *Policy and Politics,* Vol. 35 (4): 573–590.

Dell'Aringa, C. and Samek Lodovici, M. (1996) 'Policies for the Unemployed and Social Shock Absorbers: The Italian Experience', *South European Society & Politics,* Vol. 1, 172–197.

Delsen, L. (2002) 'Active Strategies for Older Workers in the Netherlands', in Jepsen, M., Foden, D. and Hutsebaut, M. (eds) *Active Strategies for Older Workers,* Brussels: European Trade Union Institute.

Department of Enterprise, Trade and Employment (2000) *Ireland Employment Action Plan 2000*: http://ec.europa.eu/employment_social/news/2000/may/nap-1999irl_en.pdf, date accessed 24/06/07.

Department of Enterprise, Trade and Employment (2002) *Ireland Employment Action Plan 2002*: http://ec.europa.eu/employment_social/news/2002/may/nap-1999irl_en.pdf, date accessed 24/06/07.

Deville, S. (2000) *France's National Retirement System*, paper for Instituto di studi economici e per l'Occupazione: new.istiseo.org/ita/dwl.php?NF=deville.pdf, date accessed 24/08/09.

Drury, E. (1993) *Age Discrimination Against Older Workers in the European Community*, London: Eurolink Age.

DTI (Department of Trade and Industry) (2005) *Equality and Diversity: Coming of Age – Consultation on the Draft Employment Equality (Age) Regulations 2006*, London: DTI.

DWP (Department of Work and Pensions) (2011) *Delivery Directorate Performance Report, Flexible New Deal Contract Employment Provision Performance*, November 2011, London: DWP. Available at: http://research.dwp.gov.uk/asd/index.php?page=ddfnd, date accessed on 21/12/11.

Ebbinghaus, B. (2001) 'When Labour and Capital Collude: The Political Economy of Early Retirement in Europe, Japan and the USA', in Ebbinghaus, B. and Manow, P. (eds) *Comparing Welfare Capitalism: Social Policy in Europe, Japan and the USA*, London: Routledge.

Ebbinghaus, B. (2006) *Reforming Early Retirement in Europe, Japan and the USA*, Oxford: Oxford University Press.

EC (European Commission) (2004) *The Social Situation in the European Union 2004*: http://epp.eurostat.ec.europa.eu/cache/ITY_OFFPUB/KE-AG-04-001/EN/KE-AG-04-001-EN.PDF, date accessed 24/06/07.

EC (2006) *Adequate and Sustainable Pensions: Technical Annex*: http://ec.europa.eu/employment_social/social_protection/docs/2006/sec_2006_304_horizontal-analysis_en.pdf, date accessed 24/06/07.

Eichhorst, W. and Rhein, T. (2005) *The European Employment Strategy and Welfare State Reform: The Case of Increased Labour Market Participation of Older Workers*, Paper presented at the 9th Conference of the EUSA, Austin, Texas.

Esping-Andersen, G. (1990) *The Three Worlds of Welfare Capitalism*, Cambridge: The Polity Press.

Esping-Anderson, G. (2000) 'The Sustainability of Welfare States into the Twenty-First Century', *International Journal of Health Services*, Vol. 30 (1): 1–12.

Estes, C. (1991) 'The New Political Economy of Ageing: Introduction and Critique', in Minkler, M. and Estes, C. (eds) *Critical Perspectives on Aging: The Political and Moral Economy of Growing Old*, New York: Baywood Publishing Ltd.

Estes, C. (1999a) 'Critical Gerontology and the New Political Economy of Ageing', in Minkler, M. and Estes, C. (eds) *Critical Gerontology: Perspectives from Political and Moral Economy*, New York: Baywood Publishing Company.

Estes, C. (1999b) 'The Aging Enterprise Revisited', in Minkler, M. and Estes, C. (eds) *Critical Gerontology: Perspectives from Political and Moral Economy*, New York: Baywood Publishing Company.

Estes, C. (2005) 'Women, Ageing and Inequality: A Feminist Perspective', in Johnson, M. (ed.) *The Cambridge Handbook of Age and Ageing*, Cambridge: Cambridge University Press.

EU Employment Observatory (2000) *Nationale Arbeitsmarktpolitiken: Basisinformationsberichte (Oesterreich)*: http://www.eu-employment-observatory.net/ersep/trd33_d/00300023.asp, date accessed 29/05/08.

European Commission: Economic and Financial Affairs (2005a) *Impact of Ageing Populations on Public Pension Expenditure*, http://ec.europa.eu/economy_finance/epc/documents/2006/ageing_spain_fiche_en.pdf, date accessed 24/06/07.

European Commission: Economic and Financial Affairs (2005b) *Revised Long-Term Projections for the Public Pension Schemes*, http://ec.europa.eu/economy_finance/publications/publication7092_en.pdf, date accessed 23/06/07.

European Communities (2000) *EU Employment and Social Policy, 1999–2001: Jobs, Cohesion, Productivity*, Employment and Social Affairs: http://ec.europa.eu/employment_social/publications/2001/ke3801681_en.pdf, date accessed 24/06/07.

European Council (2000) Presidency Conclusions: Lisbon European Council (23rd and 24th March), http://ue.eu.int/ueDocs/cms_Data/docs/pressData/en/ec/00100-r1.en0.htm, date accessed on 10/11/05.

European Foundation for the Improvement of Living and Working Conditions (2003a) *Finland: Early Retirement Pension*: http://www.eurofound.europa.eu/emire/FINLAND/ANCHOR-VARHAISEL-Auml-KEF-Ouml-RTIDSPENSION-FI.htm, date accessed 29/06/07.

European Foundation for the Improvement of Living and Working Conditions (2003b) *Portugal: Early Retirement Pension*: http://www.eurofound.europa.eu/emire/PORTUGAL/EARLYRETIREMENTSYNONYM-PT.htm, date accessed 24/06/07.

Eurostat (2007a) *Life Expectancy at Age 65, by Gender*: http://epp.eurostat.ec.europa.eu/portal/page?_pageid=1996,39140985&_dad=portal&_schema=PORTAL&screen=detailref&language=en&product=sdi_as&root=sdi_as/sdi_as/sdi_as1000, date accessed 24/06/08.

Eurostat (2007b) *Old-Age Dependency Ratio*: http://epp.eurostat.ec.europa.eu/portal/page?_pageid=1996,39140985&_dad=portal&_schema=PORTAL&screen=detail-ref&language=en&product=sdi_as&root=sdi_as/sdi_as/sdi_as_dem/sdi_as1200, date accessed 24/06/08.

Eurostat (2010) *Average Annual Income in Euros by Occupation (2007)*: http://epp.eurostat.ec.europa.eu/portal/page/portal/labour_market/earnings/main_tables

Eurostat (2011) Employment rate of older workers: epp.eurostat.ec.europa/tgm/table.doi?tab=table&init=l&language=en&pcode=tsdd-100&plugin=l, date accessed 11/11/12.

Evandrou, M. and Falkingham, J. (1995) 'Gender, Lone-Parenthood and Lifetime Income', in Falkingham, J. and Hills, J. (eds) *The Dynamic of Welfare: The Welfare State and the Life Cycle*, London: Harvester Wheatsheaf.

Evers, A. and Wolf, J. (1999) 'Political Organisation and Participation of Older People: Traditions and Changes in Five European Countries', in Walker, A. and Naegele, G. (eds) *The Politics of Old Age in Europe*, Buckingham: Open University Press.

Federal Ministry of Labour, Health and Social Affairs; Federal Ministry of Economic Affairs and Federal Ministry of Education and Cultural Affairs (1998) National Employment Plan – Austria, translated by the European Commission: http://ec.europa.eu/employment_social/employment_strategy/nap_1998/at_en.pdf, date accessed 12/06/07.

Federal Ministry for Labour, Health and Social Affairs; Federal Ministry for Economic Affairs (1999) *Implementation Report for 1999 on the National Action Plan for Employment: Austria*, European Commission: http://ec.europa.eu/employment_social/employment_strategy/99_national_en.htm, date accessed 24/06/07.

Federal Republic of Germany (1998) *National Employment Action Plan 1998*: http://ec.europa.eu/employment_social/news/1998/may/nap1998de_en.pdf, date accessed 12/06/07.

Federal Republic of Germany (1999) *National Employment Action Plan 1999*: http://ec.europa.eu/employment_social/news/1999/may/nap1999de_en.pdf, date accessed 24/06/07.

Federal Republic of Germany (2000) *National Employment Action Plan 2000*: http://www.ilo.org/public//english/employment/skills/hrdr/init/ger_1.htm, date accessed 24/06/07.

Federal Republic of Germany (2002) *National Employment Action Plan 2002*: http://www.politiquessociales.net/IMG/pdf/nap2002de_en.pdf, date accessed 01/10/12.

Federal Republic of Germany (2003) *National Employment Action Plan 2003*: http://www.einclusion-eu.org/ShowCase.asp?CaseTitleID=66&CaseID=132&MenuID=158, date accessed 24/06/07.

Federal Republic of Germany (2004) *National Employment Action Plan 2004*: http://www.politiquessociales.net/IMG/pdf/nap2004de_en.pdf, date accessed 01/10/12.

Federal Republic of Germany (2008) *Social Protection and Social Inclusion 2008–2010*: http://ec.europa.eu/social/keyDocuments.jsp?type=3&policyArea=0&subCategory=0&country=0&year=0&advSearchKey=nsr+spsi&mode=advancedSubmit&langId=en, date accessed 29/05/10.

Feroldi, D. (1998) 'Italy', in Taylor, P. (ed.) *Projects Assisting Older Workers in European Countries: A Review of the Findings of Eurolink Age*, Luxembourg: Office for Official Publications of the European Communities.

Ferrera, M. and Jessoula, M. (2005) 'Reconfiguring Italian Pensions: From Policy Stalemate to Comprehensive Reforms', in Bonoli, G. and Shinkawa, T. *Ageing and Pension Reform Around the World*, Cheltenham: Edward Elgar Publishing.

Ferrera, M. and Jessoula, M. (2007) 'Italy: A Narrow Gate for Path-Shift', in Immergut, E., Anderson, K. and Schulze, I. (eds) *The Handbook of West European Pension Politics*, Oxford: Oxford University Press.

Finnish Ministry of Labour (1998) *Finland's Employment Action Plan Based on the European Union's Employment Guidelines*: http://ec.europa.eu/employment_social/news/1998/may/nap1998fi_en.pdf, date accessed 24/06/07.

Finnish Ministry of Labour (2000) *Finland's National Action Plan for Employment April 2000 in Accordance with the EU's Employment Guidelines*: http://www.mol.fi/mol/en/99_pdf/en/90_publications/nap2000english.pdf, date accessed 24/06/07.

Finnish Ministry of Labour (2001) *Finland's National Action Plan for Employment April 2001 in Accordance with the EU's Employment Guidelines*: http://ec.europa.eu/employment_social/news/2001/may/nap2001fi_en.pdf, date accessed 24/06/07.

Finnish Ministry of Labour (2005) *Finland's National Action Plan for Employment April 2005 in Accordance with the EU's Employment Guidelines*: http://ec.europa.eu/employment_social/news/2005/may/nap2001fi_en.pdf, date accessed 24/06/07.

Fischer, F. (2003) *Reframing Public Policy: Discursive Politics and Deliberative Practices*, Oxford: Oxford University Press.

Forteza, J. and García-Zarco, B. (1998) 'Spain', in Taylor, P. (ed.) *Projects Assisting Older Workers in European Countries: A Review of the Findings of Eurolink Age*, Luxembourg: Office for Official Publications of the European Communities.

France (2000) *National Action Plan for Employment for 2000 and Results of the Plan for 1999 for France*: http://www.ispesl.it/dsl/dsl_repository/Sch34PDF08Marzo06/Sch34AboutFrenchNationalActionPlanEmploy.pdf, date accessed 24/06/07.

France (2002) *National Action Plan for Employment 2002*: http://www.centre-inffo.fr/international/spip.php?article16, date accessed 24/06/07.

France (2003) *National Action Plan for Employment for 2003 and Results of the Plan for 200 for France*: http://ec.europa.eu/employment_social/news/2002/may/nap-2002/nap2002_fr_en.pdf, date accessed 24/06/07.

France (2004) *National Action Plan for Employment 2004*: ec.europa.eu/social/ajax/BlobServlet?docId=6056&langId=en, date accessed 21/05/11.

Frerichs, F. and Naegele, G. (1998) 'Germany', in Taylor, P. (ed.) *Projects Assisting Older Workers in European Countries: A Review of the Findings of Eurolink Age*, Luxembourg: Office for Official Publications of the European Communities.

Frericks, P. and Maier, R. (2008) 'Pension Norms and Pension Reforms in Europe – The Effects on Gender Pension Gaps', *Community, Work & Family*, Vol. 11 (3): 253–271.

Gendron, B. (2011) 'Older Workers and Active Ageing in France: The Changing Early Retirement and Company Approach', *The International Journal of Human Resource Management*, Vol. 22 (6): 1221–1231.

Ginn, J. (2004) 'European Pension Privatisation: Taking Account of Gender', *Social Policy & Society*, Vol. 3 (2): 123–134.

Ginn, J. and Arber, S. (1995) 'Only Connect: Gender Relations and Ageing', in Arber, S. and Ginn, J. (eds) *Connecting Gender and Ageing: A Sociological Approach*, Buckingham: Open University Press.

Gould, R. and Saurama, L. (2004) 'From Early Exit Culture to the Policy of Active Ageing: The Case of Finland', in Maltby, T., De Vroom, B., Mirabile, M. and Øverbye, E. (eds) *Ageing and Transitions into Retirement: A Comparative Analysis of European Welfare States*, Aldershot: Ashgate.

Gourin, P. (1998) 'France', in Taylor, P. (ed.) *Projects Assisting Older Workers in European Countries: A Review of the Findings of Eurolink Age*, Luxembourg: Office for Official Publications of the European Communities.

Grand Duchy of Luxembourg (1998) *Mobilisation Founded on Continuity: National Action Plan on Employment 1998*, translated by the European Commission, http://ec.europa.eu/employment_social/news/1998/may/nap1998lu_fr.pdf, date accessed 24/06/07.

Grand Duchy of Luxembourg (1999) *National Action Plan for Employment: 1999 National Report*, http://ec.europa.eu/employment_social/news/1999/may/nap1999-lu_fr.pdf, date accessed 24/06/07.

Greek Ministry of Labour and Social Security (1998) *Social Security National Action Plan for Employment*, http://ec.europa.eu/employment_social/news/1998/may/nap1998/nap1998_el_en.pdf, date accessed 24/06/07.

Greek Ministry of Labour and Social Security (1999) *Social Security National Action Plan for Employment*, http://ec.europa.eu/employment_social/news/1999/may/nap1999/nap1999_el_en.pdf, date accessed 24/06/07.

Greek Ministry of Labour and Social Security (2000) *Social Security National Action Plan for Employment*, http://ec.europa.eu/employment_social/news/2000/may/nap-2000/nap2000_el_en.pdf, date accessed 24/06/07.

Greek Ministry of Labour and Social Security (2002) *Social Security National Action Plan for Employment*, http://ec.europa.eu/employment_social/news/2002/may/nap-2000/nap2000_el_en.pdf, date accessed 24/06/07.

Greek Ministry of Labour and Social Security (2008) *National Strategy Report on Social Protection and Social Inclusion 2008–2010*, http://ec.europa.eu/social/key-Documents.jsp?pager.offset=20&langId=en&mode=advancedSubmit&policyArea=0&subCategory=0&year=0&country=0&type=3&advSearchKey=nsr%20spsi, date accessed 27/06/09.

Grubb, D., Singh, S. and Tergeist, P. (2009) 'Activation Policies in Ireland', *OECD Social, Employment, and Migration Working Papers*, No. 75.

Guillemard, A. (1991) 'France: Massive Exit Through Unemployment Compensation', in Kohli, M., Rein, M., Guillemard, A. and Gunsteren, H. (eds) *Time for Retirement: Comparative Studies of Early Exit from the Labor Force*, Cambridge: Cambridge University Press.

Guillemard, A. (translated by Gauna, M.) (1993) 'Older Workers and the Labour Market', in Walker, A., Guillemard, A. and Alber, J. *Older People in Europe: Social and Economic Policies. The 1993 Report of the European Observatory*, Luxembourg: Commission of the European Communities.

Guillemard, A. (2001) *The Advent of a Flexible Life Course and the Reconfiguration of Welfare*, Keynote speech session II, The Impact of Welfare state policies Cost A 13 Conference 'Social Policy, Marginalisation, and Citizenship', Aalborg University, Denmark.

Guillemard, A. and Argoud, D. (2004) 'France: A Country with a Deep Early Exit Culture', in Maltby, T., de Vroom, B., Mirabile, M. L. and Øverbye, E. (eds) *Ageing and the Transition to Retirement: A Comparative Analysis of European Welfare States*, Aldershot: Ashgate.

Guillmard, A. and Jolivet, A. (2008) 'Pulling Up the Early Retirement Anchor'n', in Taylor, P (ed.) *Ageing Labour Forces Promises and Prospects*, Cheltenham: Edward Elgar.

Guillemard, A. and Rein, M. (1993) 'Comparative Patterns of Retirement: Recent Trends in Developed Societies', *Annual Review of Sociology*, Vol. 19: 469–503.

Hansen, H. (2002) 'Active Strategies for Older Workers in Denmark', in Jepsen, M., Foden, D. and Hutsebaut, M. (eds) *Active Strategies for Older Workers*, Brussels: European Trade Union Institute.

Harkness, S. and Waldfogel, J. (1999) *The Family Gap in Pay: Evidence from Seven Industrialised Countries*, CASE paper 29. London: Centre for Analysis of Social Exclusion.

Hartlapp, M. and Schmind, G. (2008) 'Labour Market Policy for "Active Ageing" in Europe: Expanding Options for Retirement Transitions', *Journal of Social Policy*, Vol. 37 (3): 409–431.

Havighurst, R., Neugarten, B. L. and Tobin, S. (1968) 'Disengagement and Patterns of Aging', in Neugarten, B. (ed.) *Middle Age and Aging*, Chicago, IL: University of Chicago Press.

Heinz, W. (2001) 'Work and the Life Course: A Cosmopolitan-Local Perspective', in Marshall, V., Heinz, W., Krüger, H. and Verma, A. (eds) *Restructuring Work and the Life Course*, Toronto: University of Toronto Press.

Hemerijck, A., Unger, B. and Visser, J. (2000) 'How Small Countries Negotiate Change. Twenty-Five Years of Policy Adjustment in Austria, the Netherlands, and Belgium', in Scharpf, F. and Schmidt, V. (eds) *Welfare and Work in the Open Economy*, Oxford: Oxford University Press.

Hills, J. (1996) 'Does Britain have a Welfare Generation?' in Walker, A. (ed.) *The New Generational Contract*, London: UCL Press Limited.

Himmelweit, S. and Land, H. (2008) *Reducing Gender Inequalities to Create a Sustainable Care System*, The Joseph Rowntree Foundation: http://www.jrf.org.uk/knowledge/findings/socialcare/pdf/2293.pdf, date accessed 13/01/09.

Hinrichs, K. (2005) 'New Century – New Paradigm: Pension Reforms in Germany', in Bonoli, G. and Shinkawa, T. *Ageing and Pension Reform Around the World*, Cheltenham: Edward Elgar Publishing.

Hirsch, D. (2003) *Crossroads After 50: Improving Choices in Work and Retirement*, The Joseph Rowntree Foundation.

Hunt, S. (2005) *The Life Course*, Basingstoke: Palgrave Macmillan.

Hytti, H. (2004) 'Early Exit from the Labour Market through the Unemployment Pathway in Finland', *European Studies*, Vol. 6 (3): 265–290.

Ilmakunnas, S. and Takala, M. (2005) *Promoting Employment Among Ageing Workers: Lessons from Successful Policy Changes in Finland*, Forthcoming in The Geneva Papers on Risk and Insurance – Issues and Practice: http://www.palgrave-journals.com/gpp/journal/v30/n4/pdf/2510049a.pdf, date accessed 24/06/07.

Immergut, E., Anderson, K. and Schulze, I. (eds) (2007) *The Handbook of West European Pension Politics*, Oxford: Oxford University Press.

Immergut, E. M. and Anderson, K. M. (2007) 'Editors' Introduction: The Dynamics of Pension Politics', in Immergut, E. M., Anderson, K. M. and Schulze, I. (eds) *The Handbook of West European Pension Politics*, Oxford: Oxford University Press, pp. 1–45.

Inglese, L. (2003) 'Early Retirement in Italy: Recent Trends', *Labour*, Vol. 17 (Special Issue): 175–207.

IPOS (International Organisation of Pension Supervisors) (2006) *Country Profile: France*, http://www.iopsweb.org/dataoecd/20/39/39628131.pdf, date accessed 24/07/08.

Italy (1999) *1999 National Action Plan for Employment*, ec.europa.eu/social/ajax/BlobServlet?docId=5810&langId=en, date accessed 24/06/07.

Italy (2004) *Italy: Employment Action Plan 2004*, http://www.lex.unict.it/eurolabor/en/documentation/altridoc/italy_nap04.pdf, date accessed 24/06/07.

Jacobi, L. and Kluve, J. (2006) *Before and After the Hartz Reforms: The Performance of Active Labour Market Policy in Germany*, IZA Discussion Paper 2100, Bonn.

Jacobs, K. and Rein, M. (1994) 'Early Retirement: Stability, Reversal, or Redefinition', in Naschold, F. and de Vroom, B. (eds) *Regulating Employment and Welfare: Company and National Policies of Labour Force Participation at the End of Worklife in Industrial Countries*, Berlin: Walter de Gruyter.

Jensen, P. (2004) 'Ageing and Work: From "Early" Exit to "Late" Exit in Denmark', in Maltby, T., De Vroom, B., Mirabile, M. and Øverbye, E. (eds) *Ageing and Transitions into Retirement: A Comparative Analysis of European Welfare States*, Aldershot: Ashgate.

Jessop, B. (1994) 'The Welfare State in Transition from Fordism to Post-Fordism', in Jessop, B., Kastendiek, H., Nielsen, K. and Pedersen, O. *The Politics of Flexibility* (2nd ed.), Aldershot: Edward Elgar Publishing Limited.

Jessop, B. (1998) 'The Rise of Governance and the Risks of Failure: The Case of Economic Development', *International Social Science Journal*, Vol. 50 (155): 29–45.

Jessop, B. (1999) 'The Changing Governance of Welfare: Recent Trends in its Primary Functions, Scale, and Modes of Coordination', *Social Policy and Administration*, Vol. 33 (4): 348–359.

Jessop, B. (2002) *The Future of the Capitalist State*, Cambridge: Polity Press.

Jönsson, L., Palme, M. and Svensson, I. (2011) *Disability Insurance, Population Health and Employment in Sweden*, Discussion Paper 17054, Cambridge, MA: National Bureau of Economic Research.

Jolivet, A. (2002) 'Active Strategies for Older Workers in France', in Jepsen, M., Foden, D. and Hutsebaut, M. (eds) *Active Strategies for Older Workers*, Brussels: European Trade Union Institute.

Kalisch, D., Aman, T. and Buchele, A. (1998) *Social and Health Policies in OECD Countries: A Survey of Current Programmes and Recent Developments*, Labour Market and Social Policy – Occasional Papers No. 33, Paris: OECD.

Kangas, O. (2007) 'Labor Against Markets', in Immergut, E., Anderson, K. and Schulze, I. (eds) *The Handbook of West European Pension Politics*, Oxford: Oxford University Press.

Kathimerini (2003) *OAED Programs*: http://www.ekathimerini.com/4dcgi/_w_articles_ell_1_20/06/2003_30979, date accessed 23/08/07.

Klercq, J. (1998) 'The Netherlands', in Taylor, P. (ed.) *Projects Assisting Older Workers in European Countries: A Review of the Findings of Eurolink Age*, Luxembourg: Office for Official Publications of the European Communities.

Knijn, T. and Kremer, M. (1997) 'Gender and the Caring Dimension of Welfare State: Toward Inclusive Citizenship', *Social Politics*, Fall: 328–361.

Kohli, M. (2005) 'Generational Changes and Generational Equity', in Johnson, M. (ed.) *The Cambridge Handbook of Age and Ageing*, Cambridge: Cambridge University Press.

Koskela, E. and Uusitalo, R. (2003) *The Un-Intended Convergence: How the Finnish Unemployment Reached the European Level*, Cesifo Working Paper No. 878, Category 4: Labour Markets, February 2003: http://www.cesifo-group.de/portal/pls/portal/docs/1/1189904.PDF, date accessed 23/08/07.

Ktistakis, Y. (2005) *Executive Summary Greek Country Report on Measures to Combat Discrimination, European Network of Legal Experts in the Non-Discrimination Field*: http://ec.europa.eu/employment_social/fundamental_rights/pdf/legnet/elsum07_en.pdf, date accessed 06/03/07.

Künemund, H. and Kolland, F. (2007) 'Work and Retirement', in Bond, J., Peace, S., Dittmann-Kohli, F. and Westerhof, G. (eds) *Ageing in Society*, London: Sage Publications.

Lafoucrière, C. (2002) 'Active Strategies for Older Workers: The European Picture Today', in Jepsen, M., Foden, D. and Hutsebaut, M. (eds) *Active Strategies for Older Workers*, Brussels: European Trade Union Institute.

Larson, M. and Pedersen, P. J. (2005) *Pathways to Early Retirement in Denmark 1984–2000*, Institute for the Study of Labour Discussion Paper 1575.

Laslett, P. (1987) 'The Emergence of the Third', *Ageing and Society*, Vol. 7: 133–160.

Lassila, J. and Valkonen, T. (2006) *The Finnish Pension Reform of 2005*, Keskusteluaiheita Discussion Papers (1000): http://www.etla.fi/files/1445_Dp1000.pdf

Leisering, L. (2003) 'Government and the Life Course', in Mortimer, J. and Shanahan, M. *Handbook in the Life Course*, New York: Kluwer Academic/Plenum Publishers.

Loretto, W. and White, P. (2004) *Out of Sight and Out of Mind: The Older Unemployed and Their Search for Work*, paper presented at Work, Employment and Society Conference, UMIST, Manchester, 1–3 September.

Macnicol, J. (2004) 'The Age Discrimination in Britain: From the 1930s to the Present', *Social Policy and Society*, Vol. 3 (2): 189–195.

Mandin, L. (2004) 'Active Aging in Europe', paper prepared for the WRAMSOC workshop in Berlin, 23–24 April 2004: http://www.kent.ac.uk/wramsoc/conferencesandworkshops/conferenceinformation/berlinconferenc/activeageingineurope.pdf, date accessed 11/3/2007.

Mann, K. (2007) 'Activation, Retirement Planning and Restraining the "Third Age"', *Social Policy and Society*, Vol. 6 (3): 279–292.

Marshall, V. and Taylor, P. (2005) 'Restructuring the Lifecourse: Work and Retirement', in Johnson, M. (ed.) *The Cambridge Handbook of Age and Ageing*, Cambridge: Cambridge University Press.

Marx, K. (1973) *Grundrisse: Foundations of the Critique of Political Economy*, translated by Martin Nicolaus, New York: Vintage.

Merla, L. (2004) 'Belgium: From Early to Progressive Retirement', in Maltby, T., De Vroom, B., Mirabile, M. and Øverbye, E. (eds) *Ageing and Transitions into Retirement: A Comparative Analysis of European Welfare States*, Aldershot: Ashgate.

Mestheneos, E. (1998) 'Greece', in Taylor, P. (ed.) *Projects Assisting Older Workers in European Countries: A Review of the Findings of Eurowork Age*, Employment and European Social Fund, Luxembourg: Office for Official Publications of the European Communities.

Meyer, T. (1998) 'Retrenchment, Reproduction, Modernization: Pension Politics and the Decline of the German Breadwinner Model', *Journal of European Social Policy*, Vol. 8 (3): 195–211.

Meyer, T. and Pfau-Effinger, B. (2006) 'Gender Arrangements and Pension Systems in Britain and Germany: Tracing Change Over Five Decades', *International Journal of Ageing and Later Life*, Vol. 1 (2): 67–110.

Meyer, T., Bridgen, P. and Riedmüller, B. (eds) (2007) *Private Pensions Versus Social Inclusion? Non-State Provisions for Citizens at Risk in Europe*, Cheltenham: Edwards Elgar.

Miller, D. (1989) *Social Justice*, Oxford: Clarendon Press.

Millar, J. (1999) 'Obligations and Autonomy in Social Welfare', in Crompton, R. (ed.) *Restructuring Gender Relations and Employment: The Decline of the Male Breadwinner*, Oxford: Oxford University Press.

Milner, S. (2007) 'Paths to Retirement: The French Life-Course Regime and the Regulation of Older Workers' Employment', *French Politics*, Vol. 5: 229–252.

Ministerie van Sociale Zaken en Werkgelegenheid (2003) *National Employment Action Plan for Employment 2003: The Netherlands*: http://www.lex.unict.it/eurolabor/documentazione/altridoc/nap04/olanda.pdf, date accessed 12/06/07.

Ministerie van Sociale Zaken en Werkgelegenheid (2004) *National Employment Action Plan for Employment 2003: The Netherlands*: http://www.lex.unict.it/eurolabor/documentazione/altridoc/nap03/olanda.pdf, date accessed 12/06/07.

Ministère du Travail, des Relations sociales, de la Famille, de la Solidarité et de la Ville (2005) *Le contrat d'accompagnement dans l'emploi*: http://www.travail-solidarite.gouv.fr/informations-pratiques/fiches-pratiques/contrats-travail/contrat-accompagnement-dans-emploi-995.html, date accessed 12/06/07.

Ministry of Economic Affairs and Ministry of Labour (1998) *National Action Plan for Employment (Denmark)*: http://ec.europa.eu/employment_social/employment_strategy/nap_1998/dk_en.pdf, date accessed 12/06/07.

Ministry of Economic Affairs and Ministry of Labour (1999) *National Action Plan for Employment (Denmark)*: http://ec.europa.eu/employment_social/employment_strategy/nap_1999/dk_en.pdf, date accessed 12/06/07.

Ministry of Economic Affairs and Ministry of Labour (2002) *NAP 2002 – The Government: Denmark's National Action Plan for Employment 2002*: http://ec.europa.eu/employment_social/employment_strategy/nap_2002/dk_en.pdf, date accessed 12/06/07.

Ministry of Social Welfare and Ministry of Health and Prevention (2008) *National Report on Strategies for Social Protection and Social Inclusion 2008–2010*: http://ec.europa.eu/social/BlobServlet?docId=2543&langId=en, date accessed 21/07/11.

Mirkin, B. (1987) 'Early Retirement as a Labor Force Policy: An International Overview', *Monthly Labor Review*, March 1987.

Mirabile, M. (2004) 'Ageing and Work in Italy', in Maltby, T., De Vroom, B., Mirabile, M. and Øverbye, E. (eds) *Ageing and Transitions into Retirement: A Comparative Analysis of European Welfare States*, Aldershot: Ashgate.

MISSOC (Mutual Information System on Social Protection) (1994) *Social Protection in the Member States of the Community: Situation on July 1st 1993 and Evolution*, Brussels: Commission of the European Communities.

MISSOC (1996) *Social Protection in the Member States of the European Union: Situation on 1 July 1995 and Evolution*, Luxembourg: Office for Official Publications of the European Communities.

MISSOC (1999) *Social Protection in the Member States of the European Union: Situation on 1 July 1998 and Evolution*, Luxembourg: Office for Official Publications of the European Communities.

MISSOC (2000) *Social Protection in the Member States of the European Union: Situation on 1 July 1999 and Evolution*, Luxembourg: Office for Official Publications of the European Communities.

MISSOC (2001) *Social Protection in the Member States of the European Union: Situation on 1 July 2000 and Evolution*, Luxembourg: Office for Official Publications of the European Communities.

MISSOC (2002) *Social Protection in the Member States of the European Union: Situation on 1 July 2001 and Evolution*, Luxembourg: Office for Official Publications of the European Communities.

MISSOC (2003) *Social Protection in the Member States of the European Union: Situation on 1 July 2002 and Evolution*, Luxembourg: Office for Official Publications of the European Communities.

MISSOC (2004) *Social Protection in the Member States of the European Union: Situation on 1 July 2003 and Evolution*, Luxembourg: Office for Official Publications of the European Communities.

MISSOC (2005) *Social Protection in the Member States of the European Union: Situation on 1 July 2004 and Evolution*, Luxembourg: Office for Official Publications of the European Communities.

MISSOC (2006) *Social Protection in the Member States of the European Union: Situation on 1 January 2006 and Evolution*, Luxembourg: Office for Official Publications of the European Communities.

MISSOC (2007) *Social Protection in the Member States of the European Union: Situation on 1 July 2006 and Evolution*, Luxembourg: Office for Official Publications of the European Communities.

MISSOC (2008) *Social Protection in the Member States of the European Union: Situation on 1 July 2007 and Evolution*, Luxembourg: Office for Official Publications of the European Communities.

MISSPEU (Mutual Information System on Social Protection in the European Union) (2001) *Old Age in Europe*, Luxembourg: Office for the Official Publications of the European Communities.

Nazroo, J. (2006) 'Ethnicity and Old Age', in Vincent, J., Phillipson, C. and Downs, M. (eds) *The Futures of Old Age*, London: Sage Publications.

Ney, S. (2005) 'Active Ageing Policies in Europe: Between Path Dependency and Path Departure', *Ageing International*, Vol. 30 (4): 325–342.

OECD (Organisation for Economic Cooperation and Development) (1997) *Ageing Working Papers – Maintaining Prosperity in an Ageing Society: The OECD Study on the Policy Implications of Ageing. Working Paper AWP 3.4: Retirement Income Systems – The Reform Process Across OECD Countries*: http://www.oecd.org/dataoecd/20/58/2429570.pdf, date accessed 25/06/07.

OECD (2000) *Reforms for an Ageing Society*: http://213.253.134.43/oecd/pdfs/browseit/8100081E.PDF, date accessed 25/06/07.

OECD (2001) *Country Chapter – Benefits and Wages: Germany* 2001: http://www.oecd.org/dataoecd/23/32/34006630.pdf, date accessed 25/06/07.

OECD (2005a) *Ageing and Employment Policies: Austria*, Paris, OECD: http://213.253.134.43 oecd/pdfs/browseit/8105121E.PDF, date accessed 25/06/07.

OECD (2005b) *Ageing and Employment Policies: Belgium*, Paris, OECD: http://213.253.134.43/oecd/pdfs/browseit/8105121E.PDF, date accessed 25/06/07.

OECD (2005c) *Ageing and Employment Policies: Denmark*, Paris, OECD: http://213.253.134.43/oecd/pdfs/browseit/8105121E.PDF, date accessed 25/06/07.

OECD (2005d) *Ageing and Employment Policies: Finland*, Paris, OECD: http://213.253.134.43/oecd/pdfs/browseit/8105121E.PDF, date accessed 25/06/07.

OECD (2005e) *Ageing and Employment Policies: France*, Paris, OECD: http://213.253.134.43/oecd/pdfs/browseit/8105121E.PDF, date accessed 25/06/07.

OECD (2005f) *Ageing and Employment Policies: Germany*, Paris, OECD: http://213.253.134.43/oecd/pdfs/browseit/8105121E.PDF, date accessed 25/06/07.

OECD (2005g) *Ageing and Employment Policies: Ireland*, Paris, OECD: http://213.253.134.43/oecd/pdfs/browseit/8106031E.PDF, date accessed 25/06/07.

OECD (2005h) *Ageing and Employment Policies: Italy*, Paris, OECD: http://213.253.134.43/oecd/pdfs/browseit/8105121E.PDF, date accessed 25/06/07.

OECD (2005i) *Ageing and Employment Policies: Luxembourg*, Paris, OECD: http://213.253.134.43/oecd/pdfs/browseit/8105121E.PDF, date accessed 25/06/07.

OECD (2005j) *Ageing and Employment Policies: Netherlands*, Paris, OECD: http://213.253.134.43/oecd/pdfs/browseit/8105121E.PDF, date accessed 25/06/07.

OECD (2005k) *Ageing and Employment Policies: Portugal*, Paris, OECD: http://213.253.134.43/oecd/pdfs/browseit/8105121E.PDF, date accessed 25/06/07.

OECD (2005l) *Ageing and Employment Policies: Spain*, Paris, OECD: http://213.253.134.43/oecd/pdfs/browseit/8105121E.PDF, date accessed 25/06/07.

OECD (2005m) *Ageing and Employment Policies: Sweden*, Paris, OECD: http://213.253.134.43/oecd/pdfs/browseit/8105121E.PDF, date accessed 25/06/07.

OECD (2005n) *Ageing and Employment Policies: United Kingdom*, Paris, OECD: www.oecd-ilibrary.org/employment/ageing-and-employment-policies-viellisse-ment-et-politiques-de-l-emploi_19901011, date accessed 11/11/12.

OECD (2006) *Ageing and Employment Policies: Live Longer, Work Longer*, Paris: OECD Publishing.

OECD (2007) *Pensions at a Glance: Public Policies Across OECD Countries*: http://mpra. ub.uni-muenchen.de/16349/ date accessed 23/08/09.

OECD (2011) *Pensions at a Glance 2011: Retirement-Income Systems in OECD and G20 Countries*: www.oecd.org/els/social/pensions/PAG

Orwell, G. (2008) *Animal Farm: A Fairy Story*, London: Penguin Books.

Paggiaro, A. and Trivellato, U. (2002) 'Assessing the Effects of the "Mobility Lists" Programme by Flexible Duration Models', *Labour*, Vol. 16: 235–266.

Palme, M. and Svensson, I. (1999) 'Social Security, Occupational Pensions, and Retirement in Sweden', in Gruber, J. and Wise, D. *Social Security and Retirement Around the World*, Chicago: Chicago University Press.

Papadopoulos, T. (2005) *The Recommodification of European Labour: Theoretical and Empirical Explorations*, Bath: European Research Institute Working Paper Series.

Paulli, A. and Tagliabue, M. (2002) 'Active Strategies for Older Workers in Italy', in Jepsen, M., Foden, D. and Hutsebaut, M. (eds) *Active Strategies for Older Workers*, Brussels: European Trade Union Institute.

Pearson, M. (1996) *Experience, Skill and Competitiveness: The Implications of an Ageing Population for the Workplace*, Dublin: European Foundation for the Improvement of Living and Working Conditions.

Pfau-Effinger, B. (1999) 'The Modernization of Family and Motherhood in Western Europe', in Crompton, R. (ed.) *Restructuring Gender Relations and Employment: The Decline of the Male Breadwinner*, Oxford: Oxford University Press.

Phillipson, C. (1982) *Capitalism and the Construction of Old Age*, Basingstoke: Macmillan.

Phillipson, C. (1996) 'Intergenerational Conflict and the Welfare State: American and British Perspectives', in Walker, A. (ed.) *The New Generational Contract*, London: UCL Press Limited.

Phillipson, C. (2004) 'Review Article. Older Workers and Retirement: Critical Perspectives on the Research Literature and Policy Implications', *Social Policy and Society*, Vol. 3 (2): 189–195.

Phillipson, C. (2005) 'The Political Economy of Old Age', in Johnson, M. (ed.) *The Cambridge Handbook of Age and Ageing*, Cambridge: Cambridge University Press.

Phillipson, C. and Walker, A. (1986) *Ageing and Social Policy*, Aldershot: Ashgate.

Piekkola, H. (2008) 'Nordic Policies on Active Ageing in the Labour Market and Some European Comparisons', *International Social Science Journal*, Vol. 58 (190): 545–557.

Pierson, P. (1996) 'The New Politics of the Welfare State', *World Politics*, Vol. 48 (2): 143–179.

Pierson, P. (2001) 'Coping With Permanent Austerity: Welfare State Restructuring in Affluent Democracies', in Pierson, P. (ed.) *The New Politics of the Welfare State*, Oxford: Oxford University Press.

Pierson, P. (2004) *Politics in Time: History, Institutions, and Social Analysis*, Princeton: Princeton University Press.

Polanyi, K. (1944) *The Great Transformation: The Political and Economic Origins of Our Time*, New York: Amereon House.

Portugal (2000) *National Action Plan 2000 – Portugal National Implementation Report*, http://www.lex.unict.it/eurolabor/documentazione/altridoc/nap04/portugal.pdf, date accessed 24/06/07.

Portugal (2004) *National Action Plan 2004–5 – Portugal National Implementation Report*, http://www.lex.unict.it/eurolabor/documentazione/altridoc/nap04/portugal.pdf, date accessed 24/06/07.

Portugal (2008) *National Action Plan for Employment (2008–2010)*, http://www.gep.msss.gov.pt/estudos/pne/pneen2008.pdf, date accessed 01/10/12.

PPI (Pensions Policy Institute) (2003) *State Pension Models*, http://www.pensionspolicyinstitute.org.uk/uploadeddocuments/PPI_-_State_Pension_Models_-_10_July_2003.pdf, date accessed 21/05/07.

Price, D. and Ginn, J. (2006) 'The Future of Inequalities in Retirement Income', in Vincent, J., Phillipson, C. and Downs, M. (eds) *The Futures of Old Age*, London: Sage Publications.

Public and Commercial Services Union (2011) *Work Programme: Providers and Contracting Arrangements*, presented at Work and Pensions Committee, London: House of Commons. Available at: http://www.publications.parliament.uk/pa/cm201012/cmselect/cmworpen/718/718vw01.htm, date accessed 20/03/12.

Quaid, M. (2002) *Workfare: Why Good Ideas Go Bad*, Toronto: University of Toronto Press.

Regeringskansliet (2002) *Sweden's Action Plan for Employment 2002*, Produced by the Ministry of Finance and the Ministry of Industry, Employment and Communications, Oslo.

Rein, M. and Klaus, J. (1993) 'Ageing and Employment Trends: A Comparative Analysis for OECD Countries', in Johnson, P. and Zimmermann, K. (eds) *Labour Markets in an Ageing Europe*, Cambridge: Cambridge University Press.

Republic of Austria: Federal Ministry for Economic Affairs and Labour (2001a) *Implementation Report for 2001 on the National Action Plan for Employment: Austria*, European Commission: http://ec.europa.eu/employment_social/employment_strategy/nap_2001/nap2001au_en.pdf, date accessed 12/06/07.

Republic of Austria: Federal Ministry for Economic Affairs and Labour (2002) *Implementation Report for 2002 on the National Action Plan for Employment: Austria*, European Commission: http://ec.europa.eu/employment_social/employment_strategy/nap_2002/nap_austria_en.pdf, date accessed 12/06/07.

Republic of Austria: Federal Ministry for Economics and Labour (2003) *Implementation Report for 2003 on the National Action Plan for Employment: Austria*, European Commission: http://ec.europa.eu/employment_social/employment_strategy/nap_2003/nap_aut_en.pdf, date accessed 12/06/07.

Republic of Austria: Federal Ministry for Economic Affairs and Labour (2004) *Implementation Report for 2004 on the National Action Plan for Employment: Austria*, European Commission: http://ec.europa.eu/employment_social/employment_strategy/nap_2004/nap2004au_en.pdf, date accessed 12/06/07.

Roschier Attorneys Ltd (2007) *Age Laws Around the World: Finland*, http://www.age-discrimination.info/international/world/finland/, date accessed 12/11/07.

Samorodov, A. (1999) *Ageing and Labour Markets for Older Workers*, Geneva: Employment and Training Department, International Labour Office.

Samuelson, W. (1980) *Economics*, London: McGraw-Hill.

Schmähl, W. (1998) 'Recent Developments of Pension Schemes in Germany: Present and Future Conflict', *Labour*, Vol. 12 (1): 143–168.

Schmähl, W. (1999) 'Pension Reforms in Germany: Major Topics, Decisions and Developments', in Müller, K., Ryll, A. and Wagener, H. J. (eds) *Transformation of Social Security: Pension in Central-Eastern Europe*, Heidelberg: Physica.

Schulze, I. and Moran, M. (2007) 'United Kingdom: Pension Politics in an Adversarial System', in Immergut, E., Anderson, K. and Schulze, I. (eds) *The Handbook of West European Pension Politics*, Oxford: Oxford University Press.

Schulze, I. and Schludi, M. (2007) 'Austria: From Electoral Cartels to Competitive Coalition-Building', in Immergut, E., Anderson, K. and Schulze, I. (eds) *The Handbook of West European Pension Politics*, Oxford: Oxford University Press.

Sjögren Lindquist, G. (2006) 'Late Careers and Career Exits in Sweden', in Blossfeld, H. P., Buchholz, S. and Hofäcker, D. (eds) *Globalization, Uncertainty and Late Careers in Society*, London: Routledge.

Social Security Administration (2004) *Social Security Programs Throughout the World: Europe, 2004*, http://www.ssa.gov/policy/docs/progdesc/ssptw/2004-2005/europe/ssptw04euro.pdf, date accessed 08/06/12.

Social Security Administration (2006) *Social Security Programs Throughout the World: Europe, 2006*, http://www.ssa.gov/policy/docs/progdesc/ssptw/2004-2006/europe/ssptw04euro.pdf, date accessed 08/06/12.

Social Security Administration (2008) *Social Security Programs Throughout the World: Europe, 2008*, http://www.ssa.gov/policy/docs/progdesc/ssptw/2008-2009/europe/ssptw08euro.pdf, date accessed 08/06/12.

Social Security Administration (2010) *Social Security Programs Throughout the World: Europe, 2010*, http://www.ssa.gov/policy/docs/progdesc/ssptw/2010-2011/europe/ssptw10europe.pdf, date accessed 08/06/12.

Spain (1999) *Employment Action Plan of the Kingdom of Spain 1999*: http://www.lex.unict.it/eurolabor/documentazione/stati/spagna/action_plan1999.pdf, date accessed 25/06/07.

Spain (2000) *Employment Action Plan of the Kingdom of Spain 2000*: http://www.lex.unict.it/eurolabor/documentazione/stati/spagna/action_plan2000.pdf, date accessed 25/06/07.

Spain (2001) *Employment Action Plan of the Kingdom of Spain 2001*: http://www.lex.unict.it/eurolabor/documentazione/stati/spagna/action_plan2001.pdf, date accessed 25/06/07.

Spain (2002) *Employment Action Plan of the Kingdom of Spain 2002*: http://www.lex.unict.it/eurolabor/documentazione/stati/spagna/action_plan2002.pdf, date accessed 25/06/07.

Spain (2004) *Employment Action Plan of the Kingdom of Spain 2004*: http://www.mtas.es/es/empleo/planemp/PNAEingles.pdf, date accessed 25/06/07.

Spain (2008) *National Report on Strategies for Social Protection and Social Inclusion of the Kingdom of Spain*, 2008–2010: http://www.socialprotection.eu/files_db/264/0-National_Report_on_Strategies_for_Social_Protection_and_Social_Inclusion_2008-2010.pdf, date accessed 08/06/10.

Spicker, P. (2008) *Generalising from Policy Research*, paper presented at the Social Policy Association Conference, 23–25th June 2008: Edinburgh.

Streeck, W. and Thelen, K. (2005) 'Introduction: Institutional Change in Advanced Political Economies', in Streeck, W. and Thelen, K. (eds) *Beyond Continuity: Institutional Change in Advanced Political Economies*, Oxford: Oxford University Press.

Sweden (1999) *Sweden's Action Plan for Employment, 1999*, http://www.regeringen.se/content/1/c4/36/48/ff971cf9.pdf, date accessed 01/10/12.

Taylor, P. (ed.) (1998) *Projects Assisting Older Workers in European Countries: A Review of the Findings of Eurolink Age*, Luxembourg: Office for Official Publications of the European Communities.

Taylor, P. (2005) *New Policies for Older Workers*, Bristol: The Policy Press.

Taylor, P. (2006) *Employment Initiatives for an Ageing Workforce in the EU15*, European Foundation for the Improvement of Living and Working Conditions, Luxembourg: Office for Official Publications of the European Communities.

Taylor, P. and Walker, A. (1996) 'Intergenerational Relations in Employment – The Attitudes of Employers and Older Workers', in Walker, A. (ed.) *The New Generational Contract: Intergenerational Relations, Old Age and Welfare*, London: UCL.

Teipen, C. and Kohli, M. (2004) 'Early Retirement in Germany', in Maltby, T., De Vroom, B., Mirabile, M. and Øverbye, E. (eds) *Ageing and Transitions into Retirement: A Comparative Analysis of European Welfare States*, Aldershot: Ashgate.

Tergeist, P. and Grubb, D. (2006) *Activation Strategies and the Performance of Employment Services in Germany, the Netherlands and the United Kingdom*, OECD Social, Employment and Migration Working Papers, Vol. 42.

Tikkanen, T. (1998a) 'Finland', in Taylor, P. (ed.) *Projects Assisting Older Workers in European Countries: A Review of the Findings of Eurolink Age*, Luxembourg: Office for Official Publications of the European Communities.

Tikkanen, T. (1998b) 'Nordic Countries', in Taylor, P. (ed.) *Projects Assisting Older Workers in European Countries: A Review of the Findings of Eurolink Age*, Luxembourg: Office for Official Publications of the European Communities.

Titmuss, R. (1955) 'Pension Systems and Population Change', *Political Quarterly*, Vol. 26: 152–166.

Thompson, W. (2009) *The Political Economy of Reform: Lessons from Pensions, Product Markets and Labour Markets in Ten OECD Countries*, Paris: OECD.

Torfing, J. (1999) 'Workfare with Welfare: Recent Reforms of the Danish Welfare State', *Journal of European Social Policy*, Vol. 9 (1): 5–28.

Triantafillou, P. (2007) 'Greece: Political Competition in a Majoritarian System', in Immergut, E., Anderson, K. and Schulze, I. (eds) *The Handbook of West European Pension Politics*, Oxford: Oxford University Press.

TUC (Trades Union Congress) (2005) *Labour Market Programmes*, http://www.tuc.org.uk/welfare/tuc-7814-f0.cfm, date accessed 04/07/05.

UK (United Kingdom) (1999) *United Kingdom Employment Action Plan 1999*: http://ec.europa.eu/employment_social/news/1999/may/nap2000uk_en.pdf, date accessed 25/06/07.

UK (2000) *United Kingdom Employment Action Plan 2000*: http://ec.europa.eu/employ-ment_social/news/2000/may/nap2000uk_en.pdf, date accessed 25/06/07.

UK (2001) *United Kingdom Employment Action Plan 2001*: http://ec.europa.eu/employ-ment_social/news/2001/may/nap2001uk_en.pdf, date accessed 25/06/07.

Uusitalo, R. and Koskela, E. (2003) *The Un-Intended Convergence: How the Finnish Unemployment Reached the European Level*, Helsinki: Palkansaajien Tutki-muslaitos Työpapereita (Labour Institute for Economic Research Discussion Papers).

Uusitalo, H. (2007) *Increased Labour Force Participation of Aging Workers: The Case of Finland*, Finnish Centre for Pensions, paper presented at ISSA, European Regional Meeting – Inclusion in Working Life, Oslo: 15–16 May 2007.

Vandenbroucke, G. and vander Hallen, P. (2002) 'Active Strategies for Older Workers in Belgium', in Jepsen, M. and Hutsebaut, M. (eds) *Active Strategies for Older Workers*, Brussels: ETUI.

Van Oorschot, W. and Jensen, P. (2009) 'Early Retirement Differences Between Denmark and the Netherlands: A Cross-National Comparison of Push and

Pull Factors in Two Small European Welfare States', *Journal of Aging Studies*, Vol. 23: 267–278.

Venge, P. (1998) 'Denmark', in Taylor, P. (ed.) *Projects Assisting Older Workers in European Countries: A Review of the Findings of Eurolink Age*, Luxembourg: Office for Official Publications of the European Communities.

Vinni, K. (2002) 'Active Strategies for Older Workers in Finland', in Jepsen, M., Foden, D. and Hutsebaut, M. (eds) *Active Strategies for Older Workers*, Brussels: European Trade Union Institute.

Vogt, M. (2006) *Austrian Summary Report on Older Workers and Introduction of Case Study Sectors*, Brussels: European Union, http://www.forba.at/data/downloads/file/198-FB%2013-06.pdf, date accessed 07/05/12.

Von Nordheim, F. (2004) 'Responding Well to the Challenge of an Ageing and Shrinking Workforce: European Union Policies in Support of Member State Efforts to Retain, Reinforce and Reintegrate Older Workers in Employment', *Social Policy and Society*, Vol. 3 (2): 145–153.

Wadensjö, E. (2002) 'Active Strategies for Older Workers in Sweden', in Jepsen, M., Foden, D. and Hutsebaut, M. (eds) *Active Strategies for Older Workers*, Brussels: European Trade Union Institute.

Wadensjö, E. and Sjögren, G. (2000) *Arbetslinjen för äldre i praktiken*. En rapport för Riksdagens revisorer, Stockholm: Institutet för social forskning (SOFI).

Walker, A. (1993) 'Living Standards and Way of Life', in Walker, A., Gullemard, A. and Alber, S. *Older People in Europe: Social and Economic Policies. The 1993 Report of the European Observatory*, Luxembourg: Commission of the European Communities.

Walker, A. (1996a) 'Introduction: The New Generational Contract', in Walker, A. (ed.) *The New Generational Contract*, London: UCL Press Limited.

Walker, A. (1996b) 'Intergenerational Relations and the Provision of Welfare', in Walker, A. (ed.) *The New Generational Contract*, London: UCL Press Limited.

Walker, A. (2002) 'A Strategy for Active Ageing', *International Social Security Review*, Vol. 55: 121–139.

Walker, A. and Maltby, T. (1997) *Ageing Europe*, Buckingham: Open University Press.

Walker, A. and Naegele, G. (1999) 'Introduction', in Naegele, G. and Walker, A. *The Politics of Old Age in Europe*, Buckingham: Open University Press.

Walker, A. and Phillipson, C. (1986) 'Introduction', in Phillipson, C. and Walker, A. *Ageing and Social Policy*, Aldershot: Ashgate.

White, S. (2000) 'Review Article: Social Rights and the Social Contract – Political Theory and the New Welfare Politics', *British Journal of Policy Studies*, Vol. 30: 507–532.

White, S. (2004) 'What's Wrong with Workfare?', *Journal of Applied Philosophy*, Vol. 21 (3): 271–285.

Whitehouse, E. (2007) *Pensions Panorama*, Washington DC: World Bank.

Wiklund, L. (1998) 'Sweden', in Taylor, P. (ed.) *Projects Assisting Older Workers in European Countries: A Review of the Findings of Eurolink Age*, Luxembourg: Office for Official Publications of the European Communities.

World Bank (1994) *Averting the Old Age Crisis: Policies to Protect the Old and Promote Growth*, Oxford: Oxford University Press.

World Health Organization (2002) *Active Ageing: A Policy Framework*: http:// whq-libdoc.who.int/hq/2002/WHO_NMH_NPH_02.8.pdf, date accessed 16/04/09.

Yeandle, S. (2001) 'Balancing Employment and Family Lives: Changing Life-Course and Experiences of Men and Women in the European Union', in Marshall, V., Heinz, W., Krüger, H. and Verma, A. (eds) *Restructuring Work and the Life Course*, Toronto: University of Toronto Press.

Index